THE SWAN

SHETLAND'S LEGACY OF SAIL

CONTENTS

ILLUSTRATIONS

ILLUSTRATIONS (CONTINUED)

FOREWORD

Shetland is fortunate to have the *Swan* as a living legacy of our rich maritime heritage. She is a lovely vessel. Many hundreds of young and old alike have experienced the special satisfaction which only a sail on her can provide. Others have admired her from afar as she has made her presence felt throughout the islands and further afield.

The *Swan* restoration has, I believe, along with the visiting boat exhibition from Sweden and the Tall Ships this summer (1999), rekindled a keen interest in the sea and Shetland's maritime past. The eventual culmination of this process will be the new Shetland Museum and Archives at Hay's Dock which will incorporate a strong maritime dimension. Hopefully the *Swan* will be closely associated with this development, particularly since the boat shed in which she was built still stands and will become part of the museum.

This timely volume traces the *Swan*'s long and varied career and presents her story within the context of Shetland's fishing history and the dramatic technological changes of the 20th century. The year 2000 will be the *Swan*'s 100th birthday and I feel especially privileged to have been involved in the project along with the team of dedicated enthusiasts who brought the *Swan* back to life. She will, I hope, continue to inspire and delight for many years to come.

Jimmy Moncrieff
(Chairman of the Swan Trust 1990 – 1998)
October, 1999

THE SWAN

SHETLAND'S LEGACY OF SAIL

FROM EARLIEST DAYS

Shetland is proud of its modern fleet of diesel-powered fishing vessels, equipped with the latest electronic aids to navigation and fish finding. But mechanical power is a comparatively recent introduction to the industry. Until the early 1900s Shetland fishermen depended on their strength to row or handle sails for propulsion and their instinct and inherited knowledge to locate the fish.

The history of fishing in Shetland can be traced back to Viking times, in the eighth and ninth centuries, when thousands of settlers arrived from Norway. Their success as seafarers was built on centuries of experience which led to the evolution of the longship, examples of which have been found preserved in peat bogs in Denmark and Norway.

This was a double-ended clinker built craft put together with iron nails and caulked with tarred rope. It had a single mast and square sail and was also propelled by oars.

The design of the longship made it ideal for breasting the waves of the North Sea and North Atlantic. Today, one thousand years after the arrival of the Norsemen, the small inshore boats still being built by craftsmen in Shetland bear an unmistakable resemblance to the Viking longships.

Until well into the 19th century fish was plentiful in Shetland's inshore waters, being caught from small boats like the yoals which are still in use in Dunrossness and Fair Isle in the south of the islands. The Faroese official, Christian Ployen, who visited Shetland on a fact-finding mission in 1839, described how his ship encountered large numbers of these double-ended clinker vessels fishing for cod and saithe off Sumburgh Head.

Commercial fishing had by then developed into a major industry in Shetland, largely through the efforts of German merchants from Hamburg, Bremen and Lübeck in the 16th and 17th centuries.

Far greater changes took place in the 18th century when Shetland landowners replaced the Germans as fish merchants and developed markets for salt ling in Spain, Portugal and the Baltic countries.

As the demand for ling increased, the range of the operation was extended progressively out to deeper water until eventually Shetland fishermen were working 40 miles from land. This was the "haaf" or deep sea fishery, which was to play such an important part in the economy of Shetland.

To reduce the distance travelled, operations were carried out from haaf stations as near the fishing grounds as possible, with a concentration in the North Mainland and in the islands of Unst, Yell and Fetlar. The main requirement for a haaf station was a clean stony beach where the salted fish could be spread out to dry and where the boats could be hauled ashore for safety when bad weather threatened.

The boat that developed for the haaf fishing was the six-oared sixern. She was deeper and beamier than the yoal but retained the basic shape and clinker construction of the old Viking longship. A typical sixern of the mid-19th century had a keel length of around 20 feet and an overall length of just under 30 feet. She carried a large square sail and each of the six-man crew pulled an oar when the wind was light.

Traditionally the haaf fishing lasted from 20th May to 12th August, when fine weather could be expected at sea and which left sufficient time at the end of the season for the last of the fish to be dried. But even in summer Shetland's climate can on occasions produce severe storms and practically every year men were lost at the haaf fishing. There were major disasters, the worst occurring in July 1832 when 105 men were lost.

The poor safety record of the sixern was mainly due to her being entirely open to the elements and liable to be swamped in extreme conditions. While fully decked boats were quite common in Shetland by the early 19th century they were considered unsuitable for the long line fishery of the haaf men. The main reason for their conservatism was that the fishermen were too poor to finance improvements and the lairds were unwilling to do so.

Soon a new force was to enter the industry with the rise of independent merchants who acquired small fully decked sloops as cod fishers. The impetus for this development was the opening of a lucrative market for salt cod in Spain

THE SWAN

1. *Swan* in her original two-masted dipping lug rig. Note lugsail tack led aft to assist in short-tacking. Photographer unknown, photo courtesy of J. L. Simpson.

2. A rare if indistinct photo of a herring drifter under jib, staysail, mainsail and "dandy" or mizzen. This is said to be the *Swan*, the photograph was taken by Captain A. Halcrow. Courtesy of Jim Goodlad.

3. Hay & Co. carpenters, 1903. Back, from left: Eddie Sinclair, John Leslie (pilot), Robbie Robertson (later diver), James Flaws, Robert Smith, Charles Smith. Middle row: James Leask, Magnie Ramsay (father of R. H. Ramsay, the photographer), Joseph Sinclair (former foreman), John Shewan (foreman from 1900 to 1940), (first name unknown) Malcolmson, Frankie Robertson, Sandy McMillan. Front: A. Williamson, Benjie McMillan, Jimmy Phillips. Photo and information courtesy of Tom Moncrieff.

4. Jack Shewan, boatbuilder and model maker. His model of the *Swan* proved invaluable in the restoration.

5. A detail from Jack Shewan's model of the *Swan*, now in Liverpool Museum. The trust has been unable to bring this model back to Shetland. However, detailed plans were made and photographs taken by Adrian Osler. These formed a valuable reference during restoration.

6. *Swan* leaving Symbister under sail. Photo: B. J. R. Jamieson.

THE SWAN

7. *Swan* under smack rig. Photographer unknown, photo courtesy of W. Simpson.

8. *Swan*. Photographer unknown, photo courtesy of W. Simpson.

9. A leading wind, both sails reefed. Photographer unknown, photo courtesy of J. L. Simpson.

10. Shooting drift nets on *Queen Adelaide* (LK1003), 1922. Photo: Rev B. Mclaughlan, courtesy of Shetland Museum.

11. A unique shot of the *Queen Adelaide* at work. Clearly shows bush rope leading on to capstan from for'd. Photo: Rev B. Mclaughlan, courtesy of Shetland Museum.

THE SWAN

12. The *Swan* underway. Date and photographer not known but pre-war and exhaust indicates she had her Gardner engine at this time. Photo courtesy of J. L. Simpson.

13. The *Swan*, now a motor boat, lying at winter moorings at the head of the West Voe, Skerries. Photographer unknown, photo courtesy of Edwin Tait.

14. At work on the *Swan* about 1930, from left: Josie Simpson, Challister; Johnnie Simpson, Vevoe, skipper 1905 to 1937, and his son Tammy Henry Simpson, Vevoe, skipper 1937 to 1956; young boy in background is Davie Kay, Challister. Photo: B. J. R. Jamieson, courtesy of Ina Hutchison.

15. Copy of receipt issued to Thomas H. Simpson and partners for shares in the *Swan* from Hay & Company, 1936.

Form No. 1

£30-0-0.Stg. Lerwick, 13th October, 1936.

Received from Mr Thomas H.Simpson, Vevoe, Whalsay, for hims
and partners,

the sum of Thirty pounds,

sterling being payment of three one-sixth shares of the hull
the boat "SWAN", LK.243, with certain furnishings and fittings as 'deta
in our letter to him, dated 29th September, 1936.

For HAY & COM[illegible]ERWICK) Ltd

[illegible] Director

THE SWAN

16. Jimmy Winchester aboard the *Swan*, lying in the Small Boat Harbour, Lerwick, 1951. Note the derrick on port side and that capstan has been moved for'd for seine net operations. Photo courtesy of Jimmy Winchester.

17. *Silver Spray*, *Swan* and *Reaper* lying at West Dock, Lerwick, in the 1950s. The *Reaper* is now preserved by the Anstruther Fisheries Museum. Photo: A. Flaws, courtesy of Shetland Museum.

THE SWAN

18. Pictured at Yarmouth by K. W. Kent, showing a "new" number, LK596.

19. "So we put onto the mud" – to clear a wire rope caught around the propeller. Photo: J. R. Botting.

20. "This is how she looked when she was sold" by J. R. Botting, 1964.

21. The *Swan* at York, 1982. Photo: Joe Irvine.

22. The aft cabin, photo taken in early 1960s by John R. Botting. The photo shows the v-lined cabin and bunk openings. In her early days the seat lockers were used for stowing coal for the boiler.

23. The Gardner gearbox removed by J. R. Botting (pictured sitting on the original engine beds) when he re-engined the *Swan* in the winter of 1962.

24. A sad sight. The *Swan* as discovered by Keith Parkes at Hartlepool in the late 1980s. Photo: Keith Parkes.

25. A virtual wreck lying half submerged in a Hartlepool dock. Photo: Keith Parkes.

26. The *Swan* lifted ashore at Hartlepool. Photo: Keith Parkes.

27. Keith Parkes pictured with the *Swan* as repairs are carried out to his "classic" ship. The work done was sufficient to save the vessel and allow her to return to Lerwick but was later condemned during the trust's restoration. Photo courtesy of Keith Parkes.

28. The *Swan* as bought by the Swan Trust. Mike Marshall, Cinecosse, at work on the video for "The *Swan*'s Return" from Hartlepool to Lerwick.

29. The *Swan* returns to Lerwick after an absence of 30 years. 13th April, 1991. Photo: Isobel Rendall.

30. The "return" crew, from left: Dennis Geldard, Robert Wishart, Tom Moncrieff, Willie Simpson and Allister Rendall. Engineer Geordie Sinclair, who kept the engine alive on the return trip, is hiding from the camera! Photo: Isobel Rendall.

and Portugal, leading to the exploitation of rich stocks of cod on the "home banks" south-west of Foula to the west of Shetland.

Foremost among those merchants were William Hay, his father-in-law Charles Ogilvy and the latter's sons, John and Charles, who in 1822 founded the company Hay & Ogilvys. In 1832 it became Hay & Ogilvy, with William Hay and Charles Ogilvy as partners.

From their headquarters at Freefield in Lerwick they built up a large business empire based on fishing and fish curing with branches in several parts of Shetland. In addition to sloops bought second-hand from the mainland, they built a large number of fully decked vessels, ranging from cod fishing vessels to the barque *North Briton*, the largest vessel ever to be built in Shetland.

THE HERRING FISHERY

While the inshore stocks of ling and cod were heavily exploited by Shetland fishermen, the herring shoals were still left largely to the Dutch who for centuries had fished the Shetland area, assembling in Bressay Sound each year and providing the stimulus for the founding of the town of Lerwick in the early part of the 17th century.

Eventually local merchants began to realise that Shetland was ideally placed as a centre for catching and curing herring and that the fishery could generate as much activity here as it was already doing at numerous ports in the North of Scotland.

Leaders in these deliberations were Hay & Ogilvy who set up small curing stations in several parts of Shetland and encouraged local fishermen to buy second-hand boats from the mainland while building up a sizeable fleet themselves. They also operated the Shetland Bank which made loans available for commercial enterprises.

Most of these vessels were of an improved type which incorporated the novelty, as far as Shetland was concerned, of a small deck for'ard giving shelter and safety that a sixern could not provide. They were known in Shetland as "half-deckers".

As more curing stations were established the islands catch of herring rose to 10,000 crans in 1830 and to 36,000 in 1833, while in 1834 55,000 barrels of salt herring were exported to markets in the West Indies and the continent.

In his book *Reminiscences of a Voyage to Shetland, Orkney and Scotland* Christian Ployen described the activity he saw at Lerwick towards the end of

July 1839 when the herring season was getting under way. He found it surprising that the Shetland vessels were smaller than their counterparts in Orkney and Scotland, considering the rougher seas to be encountered here.

His description of the boats is interesting:

Such a boat is from 80 to 100 barrels burden, but though alike in size, their build and rigging are extremely variable. Some of them are sharp fore and aft, some have a flat stern and broad bow, some have one mast with a large spret sail, foresail and jib, others have two masts and a big lug sail — in short there is the greatest variety. It is clear that the herring fishery, being still a new industry in Shetland, the people have not yet come to any fixed persuasion as to which is best adapted for the purpose.

It is unlikely that the reason for the great variety was as simple as this. It is more likely that he had witnessed a phenomenon which was to mark Shetland's fishing industry for the next one hundred years as Shetland fishermen bought cheaply what fishermen in other areas had to sell, as the latter replaced their herring boats with larger craft.

He was intrigued by the enclosed areas on these boats:

In the fore-part of each boat there is a half-deck, large enough to shelter a couple or so of men, and in the after part is a small locker, in which food and other small articles can be kept. Along each side of the boat runs a plank, on which the crew can step when executing necessary manoeuvres without treading on the fish they have caught. Five men is the usual crew of a herring boat.

By this time Hay & Ogilvy owned or managed a fleet of 100 half-deckers. A boat could be acquired for £120, complete with nets, and Ployen noted the case of five men in the books of Hay & Ogilvy who in three years had paid for a boat and nets and had besides earned £22 a year "all merely by working in the herring season eight or ten weeks in the year".

But Ployen did not realise that his visit to Shetland coincided with a downturn in the herring industry. The late 1830s were marked in Shetland by a series of poor harvests and poor fishings. Hay & Ogilvy was obliged to advance considerable sums of money to its fishermen, so that as early as 1838 the firm's position was far from secure. In 1840 the early herring fishing failed, while in September a severe gale caused the loss of several herring boats, some with all hands, while those that survived the gale suffered a severe loss of nets.

A run on the funds of the company and the Shetland Bank was met for a time by loans from the Royal Bank of Scotland but eventually the Royal Bank refused to make any further advances. In May 1842 William Hay went to Edinburgh to try to raise a loan, using his personal property as security, but before this could be arranged Hay & Ogilvy and the Shetland Bank were declared bankrupt with debts of £60,000 — a vast sum of money in those days.

The shock waves were felt throughout the entire economy of Shetland. Hundreds of fishermen, fish workers, coopers and carpenters were out of work in Lerwick and Scalloway, while other districts suffered to a lesser extent.

The once-booming herring fishing collapsed as fishermen, unable to obtain credit, were forced to stay ashore. A few half deckers managed to keep going, fishing for a limited market, but the majority remained in their noosts, quickly rotting away through neglect.

The entire property of Hay & Ogilvy was placed in the hands of trustees and a promising era in Shetland's economic history came to an end.

William Hay was not a man who gave up easily and he had the assistance of a large and loyal family. Before 1842 was out a new company had been formed — William Hay (junior) & Company, a partnership of William Hay's sons William and Charles. By the summer of 1843 they had a small fleet of sloops engaged to fish for cod while in August they resumed herring curing in a small way. In December 1843 William Hay received his discharge from bankruptcy and joined his sons in business under the name Hay & Company.

As Hay & Company started its climb towards the position of main commercial firm in Shetland its main activity was fishing. Heavily involved in the salt fish trade, the company undertook salting and drying of cod at numerous fishing stations between Sumburgh and Unst, buying ling, cod and saithe from the crews of yoals and sixerns while at North Roe they had their own small fleet of sixerns.

The company also built up a large fleet of cod smacks, with smaller vessels fishing the home banks from Scalloway while larger vessels fished as far away as Faroe, Iceland, Rockall and the Davis Strait. Their catches were dried on beaches in the Scalloway and Burra areas. This was part of the finest distant water fleet in Scotland as around 40 smacks between 70 feet and 80 feet long were owned by merchants in several parts of Shetland. To handle them required expert seamanship but inevitably several of them were lost during bad weather.

Hay & Company was also involved in the herring fishery which took decades to recover from its collapse in the early 1840s. It was divided into two

distinct seasons — the early fishing at Scalloway and Hamnavoe, usually from 1st June to 20th July, and the late fishing which started on the east side of Shetland about the middle of August and continued until the end of September or even later. On the West Side there was usually a break at the end of July while operations resumed in August, leading to heavy landings at times towards the end of the month.

By 1853 the fishery had assumed a pattern that was to continue for several years. Hay & Company had curing stations at Fetlar, Uyeasound, Burravoe, Skerries, Lerwick, and Cunningsburgh during the late fishery while those of Scalloway and Hamnavoe operated during both the early and late seasons.

Figures for that season give a clear indication of the low state of the industry. Hay & Company, as the major curer in Shetland, handled 4,865 crans of which 1,459 crans were landed at Hamnavoe and 854 at Scalloway. Fetlar came next with 704 crans, Burravoe had 675, Lerwick 451, Cunningsburgh 284, Skerries 254 and Uyeasound 183.

In 1857 William Hay devised what he termed "a new plan" for the fishery. Instead of having two types of boat — the sixern for the ling fishing and the larger partly decked boat, which some crews still fitted out for herring fishing in August — he recommended that fishermen should use their sixerns as herring boats following the end of the haaf fishing on 12th August. Many crews were already doing so and William Hay visualised a revival of the industry all over Shetland with a slightly larger dual purpose sixern. He even ordered a slightly larger class of sixern from Norway.

Unfortunately the plan came to nothing. Instead of expanding, as William Hay had hoped, the herring fishing declined steadily throughout the 1860s. In 1862 Hay & Company had only 952 barrels during the late season and the whole season's cure was carried in the hold of the *Janet Hay* to Stettin.

The 1868 season was even worse and Hay & Company had only 300 barrels to sell. With 130 barrels from another curer, Mr Methuen, they made up a cargo which was again shipped to Stettin. The following year the company had 500 barrels. By 1873 it was stated that the fishery was everywhere at a low ebb, especially at Cunningsburgh, where boats and gear were all worn out through the lack of replacement in previous years.

THE SECOND HERRING BOOM

The late 1870s saw the start of one of the most remarkable periods in Shetland's history with the revival of the herring fishery on a scale that was to surpass even the herring boom of the 1830s. In 1874 only 1100 barrels of herring were cured in Shetland, representing the catches of a fleet of small open boats, sixerns and half-deckers. In 1881 however, the islands' production rose to 59,000 barrels, caught by a fleet of 276 vessels most of which were owned by Shetland fishermen.

This was just the beginning, for in 1888 around 300,000 barrels of herring were exported and the fleet, including vessels from other ports in the UK, stood at 932.

The new era in Shetland's herring fishery can be traced to 1875. Disappointed with the lack of success on the part of local fishermen and envious of the prosperity that the herring fishery had brought to Orkney, Caithness and the North East of Scotland, Hay & Company persuaded the crews of 10 boats from Orkney to come north to fish for herring, using the firm's headquarters at Freefield, Lerwick, as their base.

The company agreed to pay the Orkney men £1 a cran from 1st June to 10th July, with the addition of a bounty of £5 to each vessel.

These vessels attracted a lot of attention when they arrived at Lerwick and started their fishing season. Their design emphasised the huge improvements that had taken place in the herring fishery since Shetland fishermen had acquired their half-deckers in the 1830s.

The new boats were large beamy vessels around 45 feet long with the vertical stem and stern of the Fifie, which was by then practically universal along the east coast of Scotland, having ousted the Scaffie of the Moray Firth.

Like the earlier herring boats, they were rigged as luggers but were entirely enclosed with a stout deck which provided space for a small fo'c'sle, with bunk beds affording a degree of comfort never before known in a herring boat.

The Orkney boats had a moderately successful season and returned in 1876, bringing with them a few boats and some curers from Wick. This started a connection with the herring curing industry at Caithness which at that time had an enviable reputation throughout the UK.

Another innovation that year was the arrival of three large boats from Buckie equipped with longlines, to fish for ling, cod, tusk and halibut on the grounds frequented by Shetland sixerns.

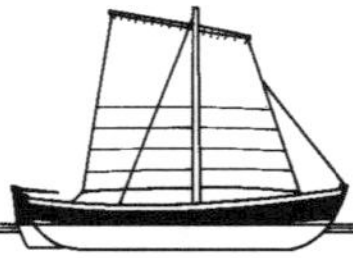

Sixern

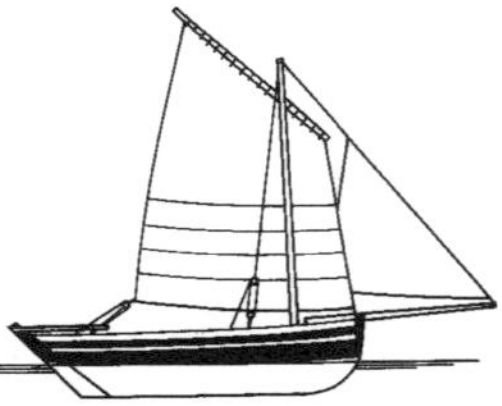

Half-decker

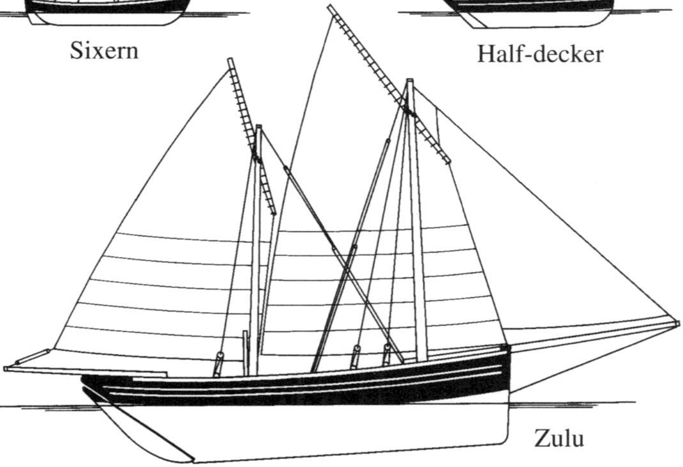

Zulu

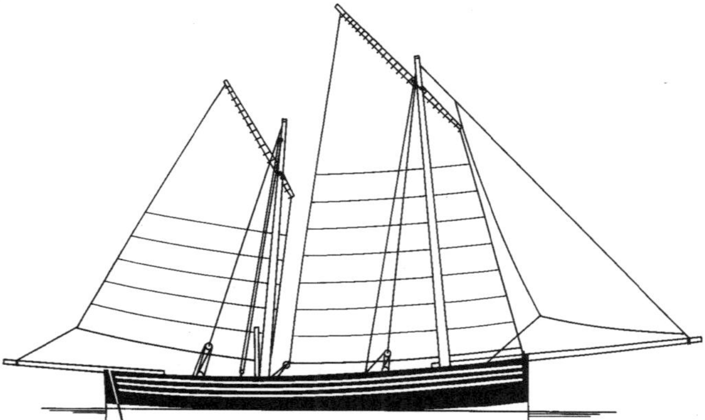

Fifie

Illustrations by Richard Stafford

The Shetlanders, accustomed to boats that could be "rowed up" to the lines while hauling, were astonished to find that the Scottish crews could not only shoot their lines but also haul them under sail by making tacks on a zig-zag course which required more skill on the part of the crew, but which, once mastered, took much of the hard work out of the job.

The three Buckie boats were instrumental in breaking the Shetlanders' aversion to big boats for the herring and long line fisheries. These vessels gave clear proof of the versatility of the new class of decked boats and their suitability for Shetland.

It was only natural that Shetland fishermen should follow the example of their counterparts in Orkney and the North-East of Scotland. In 1876 the *Defiance*, a lugger from Wick, arrived at Lerwick, having been bought by local baker, George Irvine. In February 1877, Hay & Company were approached by a crew of Burra men, led by skipper William Duncan, who were anxious to purchase a boat rather larger than the *Defiance*. The company agreed to help the men by taking a half share in the Orkney boat *Lass o' Gowrie*, which was then being offered for sale at £85. She arrived in time for the early herring fishing and from the beginning was an outstanding success.

The example of the Burra men sparked off a rush to acquire vessels of this class and, since few fishermen had sufficient capital, they had to seek the help of merchants who were willing to share the risk. In this way Hay & Company helped to finance several vessels including the *Meteor, Agnes Bruce* and *Daring*, all of which were crewed by Whalsay men. Almost all the boats that came north at this time were luggers and were converted into smacks at the earliest opportunity, because of the preference for that rig in Shetland. This is not surprising considering the long tradition of fishing in cod smacks and perhaps also in part due to the easier handling of the smack-rig in Shetland's voes.

After their conversion the long boom of the mainsail gave rise to the name generally applied to this class of vessel — the "lang boomers". In some districts they were referred to as "Buckie boats" while in Burra, because of the success of the first arrival, they were generally called "Gowries".

The demand for new boats led Hay & Company to resume the building of large vessels at Freefield. The first order was placed late in 1878 and the result was the *Hebe*, which was ready in plenty of time for the next herring season. She was followed by a long line of vessels that slid down the slipway at

Freefield - boats like the *Beaconsfield, Sunbeam* and *Lady Brassey*. In most cases the company helped to finance the vessel by taking a small share — usually one seventh.

As experience was gained by both builders and fishermen the boats became larger and better equipped and correspondingly more expensive to build. The *Beaconsfield*, launched in 1880, cost £259, while the *Matchless,* built for John Brown in 1883, cost £291, and the *Ella*, built the following year for skipper Daniel Brown of Basta in Yell, cost £322.

This chapter in the history of Hay & Company came to an end in March 1886 with the launch of the *Union*, the 34th herring boat to have been built at Freefield since the construction of decked boats began there in 1878. She had a keel length of 47 feet and a beam of 16[illegible] feet. Large by Shetland standards, she was smaller than the new generation of herring boats being launched from yards in the North-East of Scotland, giving a clear indication of a new era just starting in the herring fishery.

The lang boomers were designed for two entirely different types of fishing and proved adequate for both, in spite of their relatively small size. In March they were repainted and fitted out with long lines for the new spring fishing for ling, cod, tusk and halibut, which lasted until May or June.

Far more seaworthy than the sixern, they were on the grounds a month or two before the old-fashioned craft were launched. They carried drift nets to catch a regular supply of bait and the quality of herring caught indicated when they should switch over to the summer drift net fishery.

In those days the herring fishery was divided into two distinct parts — the early fishing from May to early July, carried on from harbours on the west side of Shetland and the North Isles, and the late fishing, which started at Lerwick in July and carried on until September or October. Then the lang boomers were hauled ashore for the winter.

Of the two fisheries the herring was by far the most important. As the number of locally owned vessels increased to around 300, the catch showed a steady increase throughout the early 1880s. As the second herring boom accelerated, Shetland's reputation attracted vessels from other ports. In 1885 800 boats fished from dozens of harbours all over Shetland, landing more than 240,000 crans that year while 370,000 barrels of salt herring were exported.

Then, as before in the history of the herring fishing, the annual catch slumped to only 31,000 crans in 1892.

Inevitably there was a marked reduction in the fleet as the older boats were scrapped or found a new role as flitboats. It was a smaller and more efficient fleet that carried through the recovery of the late 1890s when the islands' catch again reached six figures.

THE BIRTH OF THE SWAN

The small sailboats, which had started the herring boom in the 1880s, were soon outclassed by larger vessels which began to arrive in the following decade. They were from 60ft to 65ft long overall, able to put to sea in rougher conditions and carrying longer fleets of nets, with a corresponding increase in catches and profitability.

The main cause of this trend was the introduction of the steam capstan, a device which stood aft on most vessels, offset on the starboard side. It was powered by a coal-fired boiler, which in some cases was situated in the crew's quarters.

The steam capstan enabled fishermen to raise and lower heavier sails and spars thus allowing bigger boats to be handled. In conjunction with the bush rope, to which the nets were attached, it took much of the toil out of the process of hauling a fleet of drift nets.

These big boats were of two classes — the traditional Fifie of the east coast of Scotland, characterised by a vertical stem and near vertical stern post, and the Zulu, introduced about 1880, incorporating the vertical stem of the Fifie and the raking stern of an earlier class of vessel, the Scaffie.

This new chapter in Scotland's herring fishery created great activity in boatbuilding yards throughout the country. Anxious to obtain a share of this

building boom, Hay & Company resumed the construction of herring boats at Lerwick. In 1899 they put up half the cost of a new boat, the other half share to be held by one of Lerwick's top skippers, Thomas Isbister, and his partner William Watt.

The building work was supervised by David Leask who insisted that the very best material and workmanship should go into the boat. Her keel and frames were of oak, her planking pitch pine and her deck was of larch. She was of Fifie build with a keel length of 60g feet, an overall length of 67 feet and a beam of 20 feet. She was rigged as a lugger, following the pattern of fishing on the east coast of Scotland, with a large lug-sail on each of her tall masts.

The new boat was launched on Thursday, 3rd May, 1900 and even *The Shetland News*, a local newspaper, was caught up in the excitement:

An interesting event took place at Freefield Docks on Thursday when a fine new boat was launched from the yard of Messrs Hay & Company. The boat has been built to the order of Messrs Hay & Company and Mr Thomas Isbister and is acknowledged by competent judges, both local and Scotch, to be one of the finest boats afloat in the North of Scotland, as regards model, strength and workmanship.

After discussing her dimensions and mode of construction the reporter continued:

Fitted with steam capstan and all the latest labour saving appliances, the boat has every chance of a successful career, and we hope that good luck will always follow her. The launch was carried out most successfully. Miss Ottie Isbister, daughter of the skipper, performed the christening ceremony, the boat being named the Swan, and when the fastenings were cut she left the ways in good style and took to the water like a duck, being brought up in the limited space in a most masterly manner.

A large crowd witnessed the ceremony. Mr Leask the builder, is to be congratulated on this, his latest addition to the Shetland fleet.

Sadly Mr Leask did not participate in the launch of his latest creation. He became ill when the vessel was about three-quarters completed and the job was finished under the supervision of his successor John Shewan. In spite of

treatment in hospital on the mainland Mr Leask did not recover and he died on 23rd February, 1903.

It is a pity that the reporter was not interested in her colour scheme. It is known that for much of her career she was painted dark green with a red bottom and wide white cutwater, making her one of the most colourful vessels in the fleet.

Internally the *Swan* was divided into several compartments. The cabin situated aft had eight bunks, a centrally placed table with seating on both sides and a Jack Tar stove for cooking. Heating came by courtesy of the coal-fired boiler, which provided steam to drive the capstan, and coal was stored in seat lockers in the cabin.

The hold occupied most of the hull, the wings on either side divided into lockers for holding the catch and the after lockers, one on either side, held the bush rope. For'ard near the stem was a storage area for spare sails, bags of coal, etc.

On deck the most conspicious items were the two large masts between 60ft and 65ft high, their rigging and the large brown lug sails. While the after mast was fixed, the foremast could be lowered and raised by means of a hand winch. The steering wheel standing in a central position aft was connected to the rudder head by a short helm driven by a worm gear. Access to the cabin was by means of a sliding hatch. Around the hold were thick coamings about a foot high. The hold was normally kept open but could be covered with battens in bad weather.

With such large sails a boat of this size had to carry an enormous amount of ballast — more than 20 tons — for stability. This consisted of sea-rounded boulders, while below the flooring was a layer of iron ballast, although some fishermen maintained that if the heavy iron bars were placed near the keel they would create a pendulum effect which would accentuate the rolling of the vessel.

The *Swan* was given the number LK243 and was ready for the 1900 herring season. This turned out to be the most successful on record. Everything seemed to work in Shetland's favour that year. The weather was unusually settled; the shoals of dogfish which had been a plague for most of the 1890s did not appear; and since the fishing on both the east and west coasts of Scotland was well below average, Shetland curers benefited from a keen demand for salt herring. When the season ended in the middle of September the fishery officers' books recorded a total of 320,000 crans, worth £305,238. A cran was actually a

volume — not a weight — and was approximately one sixth of a ton. Herring were measured in baskets at the rate of four baskets to the cran.

Several of the local boats had earned between £800 and £900 for their season, representing catches of just under 1,000 crans, which was considered good going for a sailboat. Having cost around £600 to build it may be assumed that the *Swan* had earned a sum similar to her cost of building in her first season.

Another unusual feature of the 1900 season was the presence of a few steam drifters, the first of a new class of vessel. Some of them earned upwards of £500 in one week and the best fished vessel, the ss *Chancellor of Peterhead*, actually grossed the previously unheard of sum of £1,500 for her season.

With that kind of money being discussed around the harbours of Peterhead, Fraserburgh and Buckie, it was inevitable that many more fishermen would invest in steam drifters. Steam propulsion gave fishermen a mobility which they lacked under sail. They could go further afield in search of herring and, having hauled a shot, they could guarantee to deliver it to the curers in good condition — even on windless mornings when the sailboats were lying becalmed, their crews preparing to dump their hard-won catch back into the sea.

Such thoughts did not concern the owners of the *Swan* as she embarked on her career. They had the finest herring boat in Shetland and were determined to act as befitted their status. After a rest at the end of the summer fishery they set off for East Anglia, the first Shetland boat to take part in the autumn fishery there. The *Swan* spent six weeks at East Anglia and grossed £103 which was disappointing. Nevertheless the crew felt that they had gained valuable experience, which would stand them in good stead in the future. It was on these long journeys that the *Swan* proved herself to be a fast sailer and it is said that she once made the return trip to Lerwick in 60 hours.

In January 1902 she was one of three boats that started the winter herring fishing in Shetland. Unlike the other boats, that were either hauled ashore for

safety or anchored in a sheltered voe, these three boats were fishing at the stormiest time of the year.

The *Swan* narrowly avoided severe damage during a storm on 25th January. She was moored off Freefield at the north of Lerwick Harbour, riding by the stern to a heavy anchor, the wind blowing strong from the north-north-east. The wind suddenly increased to hurricane force and the *Swan* drove down on the quay. Fenders were quickly brought to protect the hull of the vessel and she was hauled to a more sheltered part of the quay, steam pumps working all the time to cope with the intake of water. Her ballast was removed and she was beached for repairs. It was later discovered that the anchor had broken under the strain.

Repairs occupied several weeks so that she missed part of the first winter herring season, which proved to be quite successful, albeit at a tremendous risk to boat and nets.

In those days the big sailboats were regarded as dual purpose vessels, being able to fish with longlines for white fish as well as for herring. They took part in the spring fishing, which lasted from the beginning of March until the end of May. The main species sought were ling, cod and tusk, which were salted and dried for export to the continent, and halibut which were packed with ice in large wooden boxes for shipment by the North of Scotland steamers to markets in Scotland and England.

The *Swan* was one of the last boats to take part in this fishery. About this time there was a massive increase in fishing around Shetland by trawlers from Aberdeen and other ports in the UK, which led to a big decrease in the amount of fish available for the less efficient line boats. There was also a big drop in the price of white fish because of the increased landings.

Another reason for the decline in line fishing was the greater emphasis on herring fishing as this fishery neared the peak of its profitability. This came in 1905 when no fewer than 1815 vessels took part in the summer fishery at Shetland. They landed 645,830 crans worth £576,645 and more than one million barrels of salt herring were exported, mainly to ports in the eastern Baltic.

THE SWAN GOES TO WHALSAY

Early in 1905 the *Swan* underwent a change of ownership. Thomas Isbister, ambitious as ever, decided to invest in Lerwick's first steam drifter the *Content.* With William Watt he transferred his share in the *Swan* to Hay & Company who

took in new shareholders from Whalsay while retaining a share. The new shareholders were John H. Simpson (skipper), Thomas Simpson and Robert Simpson, all from Vevoe in the north part of the island, Joseph Simpson of Challister and Laurence Hutchison of Skaw.

They took part in the spring fishing that year and on their first trip they landed 4g tons of fish, having left Lerwick on a Friday and returning the following day. Writing to John S. Nicolson, the firm's manager in Whalsay, the clerk in the company's head office described the £45 received for the catch as "a good week's work".

The new crew were to prove their skills as seamen - and also the qualities of the *Swan* - in the great gale of July 1906 which lasted for several days, causing havoc among the fishing fleets and resulting in heavy loss of life.

The late John Irvine of Lerwick — originally from Whalsay — described the experience of the crew:

My father told me many times the terrible experience the crew of the Swan had in that gale which started on 19th July, 1906. I think he told me that the gale started from the south-east to a south-west then went to a north-east.

The Swan was first on the fishing grounds. They set their nets with the wind increasing to gale force. They hauled most of their nets but left one quarter of them out. Then they swung head on, laid the mast and battened down the hatches. When the gale eased after approximately five days, there was nothing left but the bare rope. All the nets and buoys were gone. All the folk in Whalsay were mourning the loss of the Swan when the next they saw was the brave Swan coming in past the point of Fetlar with her sails reefed down. John Simpson told my father that it was a very wild and heavy sea.

It is not certain why the original owners of the *Swan* chose the lug rig in preference to the gaff-rigged ketch or "smack rig" which was far more common in Shetland. The lugsail was preferred by the Scotch men who arrived in Lerwick in the early part of this century, forming a tight-knit community while retaining their speech and the complex family naming system of places like Gardenstown and Crovie. Some of these men may have acted as advisers to skipper Thomas Isbister and crew who agreed to adopt the sail pattern of the big Scotch boats, which were making such an impression on the industry at the end of last century.

In November 1908 the skipper of the *Swan* wrote to Hay & Company suggesting that the *Swan* should be converted into a smack. John Shewan replied that this was perfectly feasible and in his opinion the mainmast would "do without alteration" while all the other spars would "make up". He estimated the cost of these alterations at £90.

The alterations involved the removal of the mizzen mast which after being reduced in size was set farther aft. The new sail plan included jib, staysail, main

sail, gaff topsail and mizzen — the typical rig of the smack, which was considered far more handy for manoeuvring in the confined spaces of Shetland voes.

She was just as successful under smack rig as she had been as a lugger. She was one of a dozen big boats based at Whalsay and she was consistently among the leaders. There were occasions when she was the top boat in the isle, as in 1913 when she grossed £711 for her summer season.

In those days neither sail boats nor steam drifter had any aids to fish finding. Their success depended on the inherited knowledge of the skipper as to where herring should be at a particular time of year, and the reports spread along the quayside as to where the best catches were coming from.

Leaving port they had only the compass to guide them, while offshore an elaborate system of landmarks, known as meids, enabled them to identify the traditional fishing grounds.

Although the herring could not be seen, the fishermen could recognise signs that indicated their presence deep below the surface. "Thick" water full of nutrients was much more promising than clear sterile water. An oily film on the sea might indicate dense shoals of herring below the surface, while seabirds diving or whales blowing often indicated herring near the surface.

When it was time to shoot the nets, the smack was turned away before the

wind, running under reduced sail or perhaps under bare poles if the wind was strong. In more severe conditions the crew could put out the heavy drogue, which always lay ready at the stern, to reduce the boat's speed. It was often used when approaching the quay. On the other hand there were times when the wind was so light that the men had to use large oars to set the nets out.

The *Swan's* fleet consisted of 60 nets, each 55 yards long, fastened together and neatly stacked in the hold ready to be shot. During this process the nets would be attached at the bottom to a long length of heavy bush rope which kept the nets together. They hung vertically in the water, suspended from large canvas bowes (buoys) which floated on the surface.

The bush rope was stored aft, coiled neatly in two after wings of the hold, half on each side of the vessel to maintain balance, while the bowes were stored for'ard in the fore part of the hold. A member of the crew made sure that the bush rope ran out clear while the cook stood for'ard in the hold, passing up the bowes to a man on deck.

When the skipper gave the signal the first bowe was flung overboard, attached to the end of the bush rope. As the rope began to run out, deft hands tied onto it the ends of the fasteners or stoppers along the bottom of each net in turn. At the same time a bowe rope was attached to the bauk or cork line of the net wherever two nets were joined together. It was a skilful operation, requiring the attention of all the crew to ensure that the nets ran out clear.

When all the nets were shot, more bush rope was paid out and made fast for'ard. As the rope tightened the boat swung round bow-on to the nets which could be seen as a long line of bowes, rising and falling in the swell as far as the eye could see.

There remained one very important job to do — one that never failed to alarm a visitor who wanted to experience a night "at the herring". The main mast had to be lowered to help the boat lie more comfortably head to wind. Also, if left standing, it would have caused enormous strain on the vessel as she rolled and pitched while lying at the nets.

Captain Halcrow described the scene in his book *The Sail Fishermen of Shetland*, as he recalled a trip on another sailboat from Whalsay, the *Gracie Brown.*

I saw the hands slackening the stays. I saw the winch handles manned by four pairs of brawny arms. The skipper shouted 'lower away'. There was a

31. After removal of engine and ballast to reduce her weight the *Swan* was lifted ashore at Holmsgarth, Lerwick. Here Billy Smith is removing the bulwarks. Photo: Brian Wishart.

32. The *Swan* is lowered gently into position and shored-up ready for restoration. Photo: Malcolm Younger.

33. As work got underway the extent of the damage and decay to her timbers became apparent. Photo: Graeme Storey.

34. Carpenter Billy Smith removing old timbers. Photo: Graeme Storey.

35. A chain saw – a luxury not available when the *Swan* was built. Modern tools make lighter work of heavy timbers as Wilfie Bruce demonstrates. Although modern tools were used where appropriate the restoration was based on traditional boatbuilding methods and skills. Photo: Graeme Storey.

36. Raymond Sinclair puts the finishing touches to the new aft stem. Old skills are the only way to ensure a perfect fit with the massive timbers used in traditional boat construction. Photo: John Martin Tulloch.

37. Some of the carpenters take a break for a photo. At this stage the reframing is largely complete and work on replanking is about to start. Back, from left: Jim Tait, Douglas Moore, Raymond Sinclair, John Martin Tulloch of the Malakoff. Front: Wilfred Bruce, Billy Smith and Gordon Smith working for the Swan Trust. Photo: Graeme Storey.

38. To fend off the worst of Shetland's winter weather scaffolding and tarpaulins formed a "tent" over the *Swan*. The new framing was fitted and fastened one at a time to the original hull planking and here the top planking is being replaced. In this way her shape was preserved. Photo: Allister Rendall.

39. Planking nearing completion. Here the original keelson can be seen between the new keel and garboard. Photo:y Allister Rendall.

40. Well through the work on the new deck beams. Photo: Allister Rendall.

41. Deck work nearing completion. The iroko planking of the poop deck is more or less complete. Note the hole for the mast for'd which allowed the mast to be lowered at sea. This was another traditional feature retained in the rebuilding. Photo: Allister Rendall.

42. All spruced up after her restoration the *Swan* returns to her natural element, 16th April, 1994. Photo: Allister Rendall.

43. The Cummins engine, refurbished by Malakoff and Moore Ltd. is lifted aboard. Photo: Jimmy Moncrieff.

44. The Liverpool Museum model of the *Swan* was helpful in planning fittings, deck layout and producing the new sail plan. Drawings were made of the model for the Swan Trust by Adrian Osler. Photo: Adrian Osler.

45. Sparmaker and rigger Tommi Nielson supervises hoisting the mainmast into position ready for rigging. Photo: Allister Rendall.

46. Tommi Nielson splicing the rigging. During freezing winter weather his "peerie" finger was dislocated and bent at right angles but it was so cold he didn't feel it! Photo: Tom Moncrieff.

47. Lashing the new mainsail to the gaff and mast hoops. Photo: Barbara Henry.

terrible banging of blocks and the swish of swinging wire and ropes. I momentarily wondered if the stout, bolted, oak trunkway would stand the strain - and the mast was down, resting horizontally in a crutch fixed to the mizzen mast.

The sail was stowed on the port side of the vessel — a universal rule in smacks and luggers alike — leaving the starboard side clear for shooting and hauling the nets.

With the nets out and the mast laid there was time for supper before all hands turned in for a few hours sleep — except for the crewmember whose turn it was to act as watchman. He had to watch for signs of worsening weather and for signs of "weight on the bowes", which meant that a dense shoal of herring had struck the nets and that the crew had better start hauling if they wanted to retrieve their costly gear. The watchman also had to stoke the boiler to ensure that there was a head of steam on the capstan. As he did so the cabin became uncomfortably hot for the six men who were trying to sleep.

The nets were left during the few hours of darkness when the herring — if they were in that area — performed their daily pattern of rising towards the surface, following the plankton on which they feed. Sometimes the apex of their rise was still below the level of the driftnets and the crew would have an easy haul, with nothing to show for their day's work.

As the crew came on deck at one or two o'clock they had no idea what to expect. The next hour would decide whether or not their efforts had been in vain.

The capstan was set in motion and the cook took up his position in the rope locker for his dreary stint, which lasted for several hours, of coiling hand over hand the miles of cold clammy and tarry bush rope.

The capstan merely hauled the boat up to the nets. Each net still had to be hauled aboard by muscle power alone involving team work by all crew members. With the boat lying stern on to the fleet, the nets were hauled over the side. The man standing furthest aft hauled the sole bauk and untied the stoppers from the bush rope as it passed down to the man in the rope locker. Another man farther for'ard hauled the upper bauk, untied the bowes and passed then down to the hold. The remaining four men stood on a platform in the hold, known as the shakkin planks, hauling in the lint in a pile which got higher as the nets were pulled in.

The nets were hauled over a roller fixed to the hatch coaming and the herring caught in the meshes were shaken out vigorously, to fall in a silver cascade on the deck. Circular lids incorporated in the deck were removed to let the herring fall into the lockers in the hold, with a system of chutes which allowed herring to pass into the lockers on the port side of the vessel to maintain its trim.

When the nets had all been hauled, the foremast was raised, the sails were set and the *Swan* headed for port, the crew hoping that the wind would hold steady so that they could reach the market and discharge their catch in good condition.

If the wind failed, the *Swan,* in common with all sailboats, faced having to dump her catch of herring back into the sea as the boat lay becalmed. They had no choice if they were to try their luck again the following night. There was no demand for overday herring which lost their freshness very quickly, making them unsuitable for curing.

While Lerwick rose quickly in importance after 1905 to become the premier herring port in Shetland and one of the main herring ports in Europe, many smaller ports were left with little more than crumbling jetties to remind them of the herring boom. Whalsay was one of the very few which maintained its activity with a thriving curing station at Symbister. Inevitably the *Swan's* crew chose their home port whenever possible, unless Lerwick was much nearer. Discharging at Symbister on a Saturday morning the crew had a longer weekend at home. They did not leave again until Monday afternoon since there was no fishing on Sunday, and no market on Monday either, to ensure that the Sabbath was observed.

On the way ashore the crew had to haul the nets up from the hold onto the deck to remove the herring which were too firmly enmeshed to be shaken out when the nets came aboard. Moreover, the shakkin planks had to be removed so that the catch could be discharged.

At Lerwick the herring were sold by auction at the fish market, the price based on the quality of a sample in a basket. Each curing firm had its own station by the shore, with its own private jetty where boats could unload their catches.

Discharging a catch of herring was a laborious job, since the sailboats did not have derricks to swing the basket ashore. A rope was slung from the top of the mizzen mast to the mid-point of the mainmast with a block hanging from it directly over the hold. The loaded baskets were hauled up by means of a small

drum on the capstan and pushed or hauled onto the jetty as the rope was paid out.

The herring were tipped into a bogie, which ran on rails right up to the wooden farlins where the women gutted and packed the herring in barrels with salt for export to the Continent. With the catch discharged the boat's crew had to lay the nets back in the hold ready to be shot again. And when that was completed the crew had time for a meal and a few hours sleep before casting off for another fishing trip.

There was a variation on this theme at the weekends when, after discharging the catch, the mast was lowered and the nets were draped over it to allow the wind to dry them and prevent them from "takkin heat," which would have happened if they had been left in a pile on deck or in the hold. From time to time they were cutched — dipped in a tank in a solution of cutch, an extract of plants from the Far East, used as a preservative. Occasionally the nets would be taken ashore by horse and cart and spread out to dry in the North Park at Symbister.

The herring fishing came to an end at the beginning of September and there was a long break before the next summer season in May. Under her Whalsay crew she no longer fished at East Anglia nor did she take part in the winter herring. The spring fishing too became uneconomic due to a massive increase in the effort by trawlers from Aberdeen and other ports in the UK.

During the winter months many of Shetland's herring fishermen turned to a different kind of fishing, setting baited lines to catch haddock within a few miles of the shore from small open haddock boats. The men from Vevoe were able to crew two boats at the beginning of this century. The third strand in their economy was their crofts, which provided vegetables, meat and milk and a small amount of cash from the sale of livestock.

From September to May the islands' herring fleet lay secure from the winter storms in secluded, sheltered voes. While the North and South Voes at Symbister are safe from most wind directions, they are exposed to winds from a westerly direction, so the herring boats were anchored in more sheltered places like Catfirth on the Mainland, or Burravoe in the south of Yell. The owners of the *Swan* chose the West Voe of the neighbouring isles of Skerries, knowing that the vessel would lie there safely no matter how strong the wind might blow.

The *Swan* was to lie there for four years during the First World War. A large number of Whalsay men were called up for service, leaving insufficient men to

form crews for the entire fleet. Some of the *Swan's* crew served on the *Gracie Brown*, one of the few boats that managed to keep going throughout the war.

Several of the boats that had lain idle or had been hauled ashore had suffered from neglect and had to be scrapped. The *Swan*, however, had been carefully tended, being pumped out regularly by a family in Skerries.

MOTOR POWER

In the 1920s the *Swan* was one of the steadily-declining fleet of sailboats which found it hard to compete with the steam drifters. These were by then dominant in the industry with around 300 taking part in the summer fishery at Lerwick each year. Catches were high and market prices were low and the small catches taken by the sailboats gave at best a very modest income, in spite of their low running costs.

By this time several of the big Zulus and Fifies had been converted to motor, with the installation of paraffin engines. In the 1930s, the price of steam coal rose considerably while the price of herring fell even lower due to currency problems in Germany, which by then was the main market for Shetland herring. As a result the motor boats were more profitable than the majority of the steam drifters.

In 1935, for example, when 304 vessels landed 203,960 crans at Shetland ports, local drifters averaged £1150 for the season, motor boats £1000, and sailboats £500. It is interesting to note that Whalsay was still the second most important herring port with landings of 4104 crans but a long way behind Lerwick which handled 190,015 crans.

In September 1935 when the *Swan* was one of only five sailboats left in Shetland the working owners approached Hay & Company with a suggestion that an engine should be installed in the vessel. Hay & Company replied that they could not agree to this expense but that they would be willing to sell their share of the vessel to the partners in Whalsay. The offer was accepted and the *Swan* became wholly owned by Whalsay men.

While a new engine was beyond their means, they saw an opportunity a year later when the old flitboat *Geirhilda* was offered for sale. They bought the vessel, a former fishing boat, and after removing her 75 hp Gardner engine they offered the hull for sale.

The *Swan* was taken to Lerwick for extensive alterations. While many old vessels had the engine installed between the cabin and the hold, with the reverse gear actually in the cabin, the owners of the *Swan* decided to utilise a space near

the bows which became available once the mainmast was removed. Inevitably this required a long shaft passing almost the entire length of the vessel to the propeller. One of the most difficult parts of the job was to bore through the after stem for the stern tube.

At the same time it was decided to move the steam capstan to the foredeck. The space between the hold and engine room was divided into two compartments — the boiler housed on the port side and a new rope locker on the starboard side. As a result of these alterations the crew were given more room in the cabin and they also obtained the full use of the hold for carrying herring, since the two aft lockers were no longer required for holding the bush rope.

On deck the mainmast trunk was planked over and in place of the tall mast came a much shorter one stepped on the deck. This carried a swinging derrick for discharging the catch.

Alterations at the stern involved repositioning the steering wheel and the construction of a wheelhouse. This was merely a hut-like structure with a seat for the helmsman while the compass swung in a binnacle overhead. An improved steering system was introduced with chains connected to the rudder-head. Finally, the old mizzen mast was removed, freeing the cabin from this obstruction. In its place came a much smaller mast stepped onto the deck abaft the wheelhouse.

These alterations were carried out by a Lerwick firm of carpenters. The engine was installed by George Gair while the final adjustments were made by Mr Jack Moore of Scalloway, the recognised expert on all matters affecting Gardner engines.

Another important change at this time was the retirement of her skipper John Henry Simpson at the age of 67. He was succeeded by his son Thomas Henry Simpson. New partners in the ownership of the vessel following the withdrawal of Hay & Company were Robert J. Hutchison and Joseph Kay.

It must have been a strange feeling for a crew accustomed to the restraints imposed by sail to be able to steer the vessel in whatever direction they wanted — even straight into the wind!

With their new-found mobility and greater speed they were able to get to and

from the grounds more quickly and were able to use the extra time by working a greater number of nets — 70 as opposed to 60. They also engaged an additional crewmember so that all eight berths in the cabin were required.

The for'ard position of the engine posed problems for the "driver", as the engineer of a Shetland fishing vessel was always known at that time. He had to be ready to respond to the skipper's command at a moment's notice. A large gear lever protruded through the deck so that he could select ahead, neutral or astern while the engine was running at a slow speed, but to increase or reduce speed he had to get down the ladder and alter the controls by hand.

An instant start from cold was impossible. Each of the four cylinders had to be heated by a lamp attached to the inlet manifold for about 20 minutes so that the paraffin could vaporise. Once this was done the engine would start with the handle and give no more trouble during the day, if the driver was doing his job as regards maintenance. It is said that on one occasion, when the *Swan* had been lying at anchor for more than seven months, the lamps were applied as usual whereupon the engine started with the first swing of the handle.

The extra weight for'ard was balanced by more ballast aft to ensure that the vessel was in trim. A large number of former sailboats had 75hp Gardner engines installed by this time and it was found that the *Swan* was the fastest of them all.

The only cause for regret was that during her conversion to motor a semi-circular section had to be cut out of the rudder to accommodate her propeller and this made her far less responsive to the helm than when she had been under sail.

During the Second World War the *Swan* was called up by the Admiralty for harbour duties. A number of Shetland boats went south at this time, some never to return. A Navy crew were given the job of taking these boats to the mainland. They were stumped by the *Swan* however, since none of them could handle the complexities of the 75hp Gardner. This was solved when Mr Laurence Irvine of Whalsay agreed to take her down to Fraserburgh.

Renamed *Rivulet*, the *Swan* was stationed at the Tail of the Bank in the Clyde estuary, her job being to collect waste paper from the big ships, including the ss *Queen Mary*, that assembled there for Atlantic convoys. Mr William Leask, a former Whalsay man now who moved to Lerwick, recaled the day while he was serving on a minesweeper that he stepped onto her deck while she was berthed at Albert Dock in Greenock. He admitted that it was a nostalgic

moment that brought back memories of happier days. Another Shetlander, John William Sutherland of Skerries, was engineer on the *Swan* for a considerable part of this period.

The *Swan* returned from the war in 1945 little more than a bare hull and badly neglected. She was laid at her usual moorings in the West Voe of Skerries while her owners decided what to do with her. She was then taken to Lerwick, where she was berthed at the West Dock while extensive repairs were carried out. There were more changes in her ownership at this time.

For three years the *Swan* followed her usual pattern of fishing for herring during the summer, then spending the rest of the year lying idly at anchor. However, there was another change in store as she was equipped to catch haddock and whiting.

During the war the seine net had replace the old-fashioned haddock line and it was discovered that any boat with an engine could increase its earning capacity by having a seine net winch installed, enabling it to fish profitably throughout the year.

It was soon apparent that the big herring boats made ideal seine net vessels and many of them became dual purpose vessels, fishing with driftnets for herring during the summer months then switching to seine netting during the rest of the year.

During the winter of 1952/53 a seine net winch was installed in the *Swan* and the next few years were perhaps the busiest of her entire life. Although her engine was of modest power she towed a seine net with the best of them. Fish was plentiful in those days and her main fishing grounds were those on the east side of Shetland, especially the Fetlar Firth, where she had many a haul of 60 or 70 boxes, towing a net with only nine coils of rope on each side. A slight modification was made to the engine controls so that her speed could be regulated from the wheelhouse.

There was little demand for white fish in Shetland at that time and like most Whalsay boats the *Swan's* crew had to consign their fish to Aberdeen for sale. The catch was gutted and packed with ice in boxes which were landed at Victoria Pier at Lerwick and transferred to the *St Clair* on Tuesdays and Saturdays, which came to be known as "shipping days".

Even at Aberdeen prices were low, while freight charges were relatively high, and the crew considered themselves lucky when they heard that the catch had cleared £1 per box.

While home at Whalsay for the weekend, or when bad weather prevented fishing, the *Swan* was anchored in the North Voe at Symbister, one of several seine netters which did so, each crew using a "drag boat" to make the passage between the fishing boats and the shore. This could be a hazardous undertaking when the wind was blowing strong from a westerly direction.

Willie Simpson, who is now a member of the Swan Trust, joined the *Swan* for a short period in 1953, becoming a full-time crew member the following year. His father, Thomas Henry Simpson, was still skipper and the other shareholders were Robert J. Hutchison, David Kay, Alex Hutchison, John Johnson and Laurence Bruce. Laurence Shearer and Samuel Irvine completed the crew. With nine men forming the crew for the seine net fishing an additional bunk had to be installed in the cabin.

Willie recalls the occasion when the *Swan* almost became a total loss. Coming in from the herring fishing they were approaching Lerwick in thick fog. In those days fishing boats were not equipped with radar. The only aids to navigation were the compass and constant watchfulness.

Willie was sharing the watch with Alex Hutchison who was in the wheelhouse steering the vessel. Willie was just going below to waken the rest of the crew when Alex reduced speed and shouted in desperation that they were heading for a cliff. Willie raced for'ard, grabbed the gear lever, got the engine in reverse and braced himself for the inevitable collision.

By a strange stroke of luck the *Swan* was heading for a large opening in the sheer cliff face known as the Cave of the Bard and while she hit the cliff face it was only a glancing blow. The *Swan* was out of action for several days getting a new stem fitted on the Malakoff slipway in Lerwick.

The 1950s were years of expansion for the Shetland fishing fleet with new boats being added at the rate of several each year, replacing old boats which were being scrapped at the same rate. The new boats were dual purpose vessels designed for both drift net and seine net fishing with engines up to 150hp. They

could tow a seine net on rougher bottom while the crew's quarters were of a standard never seen before in inshore fishing vessels.

In 1956 the *Swan* took part in the herring fishing for the last time. The following year five of her crew, Robert J. Hutchison (skipper), Alex Hutchison, David Kay, Laurence Shearer and Willie Simpson took delivery of their new vessel, the 55 foot long *Brighter Morn*. Samuel Irvine, who had been on the *Swan* for two years, took delivery of the 70 foot long *Orion* the same year. Thomas Henry Simpson, who had been skipper of the *Swan* since 1937, went as a crew member on other boats for a few years and then retired.

The *Swan* was anchored at Burravoe in Yell where she lay for over a year. Several boats of her age had been broken up to make fencing posts but she avoided this fate when she was bought by Lerwick fish merchant John Phillips, who realised that she was far too important a vessel to be broken up. He bought her without any definite plans for her future use.

She was towed to Lerwick by the *Brighter Morn* and although the engine had not been running since her last fishing season, her former crew had it running on the way to Lerwick.

The *Swan* was berthed at the West Dock where she was to spend the next three years while Mr Phillips considered how his acquisition could best be utilised.

THE LONG ROAD TO RESTORATION

Eventually the *Swan* atttracted the attention of an English shipowner, Mr W. J. Havens, who specialised in the restoration of old shipping vessels with a view to re-sale.

He arrived in Shetland at the end of July 1960 and bought three old fishing boats which he considered worthy of restoration - the *Swan* and *Silver Spray*, both lying at Lerwick, and the *Speedwell* which was berthed at Scalloway.

His motives were outlined in an article in *The Shetland News* of 4th August, 1960:

Mr Havens explained that believing there is a great tradition embodied in them and that specimens of all kinds should be preserved, he buys them and restores them to their original condition. As he cannot afford himself to build up a museum of this kind, he re-sells them to people who profess to have the intention of keeping them in their restored condition. He has already done some Brixham trawlers, Bristol Channel sailing cutters, Fifie smacks and Zulus.

Mr Havens' effort is therefore in line with the Shetland Folk Society's recent purchase of a sixern for restoration of which he was very interested to learn. He made the comment that fishing had obviously been the lifeblood of Shetland so that nothing was so much worth preserving as specimens of the fishing craft in the islands.

Mr Havens said he was particularly surprised that local effort had not been made to buy the Swan from her last owner for her hull was so sound. Restored to her old form, and with her ample deck space, she could, he said, have been an attraction for tourists to be shown over her for 6d at a time; and also be taken out for trips in the harbour under sail at so much a time. The income thus gained would he said in time have paid for the preservation, and would have given employment to an old skipper and one or two men.

Mr Havens had arrived with the former Admiralty salvage vessel MFV 1134 and on Friday 28th July he supervised her departure with both the *Swan* and *Silver Spray* in tow. Off Bressay lighthouse the fishing boats broke adrift causing some damage to the *Silver Spray*. All three vessels returned to Lerwick and the salvage vessel sailed for Grimsby the following Sunday, towing the *Swan* alone. In his editorial Dr T. M. Y. Manson of *The Shetland News* expressed his delight that the *Swan* was to be restored:

There is nothing that calls so much sentiment as a ship or boat that has had her day and is about to be disposed of or broken up. This was illustrated quite recently when it became known that the one time outstanding sail herring boat Swan had been purchased for removal south.

Today we publish an article which comes as a pleasant surprise, for it shows that the Swan will be restored to her original condition as a sail boat, in common with a number of old-time craft.

The Swan however will then be put up for sale: there are always some people interested in preserving such craft.

It is not too much to suggest that a local effort should be made in Shetland to buy back the Swan and keep her along with the sixern recently bought by the Folk Society. Both could keep alive the memory of their eras and be of lasting interest both to local people and to tourists.

Indifference to such matters is never a healthy sign. The most vigorous and progressive peoples are always conscious of their past.

Mr Havens' plans for restoring the *Swan* did not materialise. She was sold again and her new owners carried out alterations before they too gave up on the job. Her next owner was Mr John R. Botting who realised that the *Swan* was still useful as a fishing boat and he set about the task of putting right some of the defects caused by the earlier attempt at conversion.

Mr Botting described his first impressions on seeing the old vessel:

I first saw the Swan late in 1960. She was moored in the old docks of Grimsby, looking very sorry for herself. It took me a long time to find out who her owners were. After a long debate I paid £200 for her. She looked a right mess and was very wet below the decks owing to the day lights that had been cut into her decks. Her decks also required caulking. Removal of the rain water and the mess of wood and hardwood was my first priority.

While the decks were being recaulked, Mr Botting had a two and a half feet high bulwark and oak rail fitted for safety reasons and to support a pot hauler for working crab creels.

He did not like the old Gardner engine which he found heavy to turn over by himself. In his opinion: "it was a work of art that one did not suffer damage to oneself when it back fired, as it was prone to do."

The removal of the engine occupied several weekends since Mr Botting was working as a truck driver during the week and having to drive 160 miles to Grimsby, then driving back to London to resume working on Monday morning.

After the engine was removed Mr Botting had to remove about 50 feet of propeller shaft and a further 50 feet of exhaust pipe which still ran along the deck on the port side. Then with his brother's help he fitted bearers in a new engine room for'ard of the cabin to support a Hudson Invader 168 engine delivering 150 hp at 1700 rpm. With the new engine installed the *Swan* was again ready for fishing, the engine driving her at a speed of about eight knots. She was given the registration number LK596 since the number LK243 had been transferred to the Whalsay boat *Dewy Rose*.

Unfortunately the engine only had a short life in the *Swan* , as Mr Botting explains: "The winter of 1962-1963 was one of the worst since 1947. It was so cold with snow and ice. I was at home in London because my wife was giving birth to twin girls. When I returned to Grimsby I was told that the engine had frozen solid and split the engine block. The crew had forgot to drain the water out. I find fishermen in those days were not used to modern engines."

Undaunted, Mr Botting began to look for a replacement and found a 150 hp Caterpillar engine and a two-to-one reduction oil-operated gearbox made by Hindmarch Wheel Drive Ltd. Mr Botting had to instruct the engineer as to how the engine and gearbox should be operated. The *Swan* was put on the slip at Grimsby to have the engine and gearbox fitted and at the same time a new keel strap and bow strap were fitted.

While fishing from Grimsby the *Swan* had a crew of five. Her skipper Stan Upcroft was an experienced fisherman who knew the North Sea well. Although the price of crabs was often poor, owner and crew made a fairly good living.

In 1964 Mr Botting agreed to sell the *Swan* to a friend Joe Davis, who had a diving business. For the next few years information about her movements is scanty.

In 1973 a group of enthusiasts in Lerwick formed the Shetland Maritime Trust, the aim being to preserve as much as possible of Shetland's maritime heritage. Its chairman was Harry Jamieson and its secretary Ivor Hawkins. The group had learnt that the *Swan* was berthed at Alexandra Dock in Grimsby and apparently in quite good condition.

The trust sought the help of J. L. Richardson & Sons Ltd, a firm of maritime surveyors in Grimsby, in finding out more about the *Swan's* condition and whether she could stand the journey back to Shetland.

An employee of the firm T. W. S. Chambers carried out a detailed survey of the vessel and his report was far from encouraging. He found out all the structures above the deck — masts, wheelhouse, bulwarks and stanchions — had been removed and that her interior had been stripped as her owners converted her into a houseboat.

He was unable to gain access to the for'ard part of the vessel since the doors were locked and the keys in the possession of other prospective purchasers with whom the owners could not make contact at that time. As a result inspection of the hull was confined to the engine room. This was sufficient to confirm that the main frames were generally rotten at their tops and that some of the upper planks were rotten or missing. It was clear that repair work was in progress "apparently on a part-time amateur basis." She still had the six cylinder Caterpillar engine.

Mr Chambers informed the trust that the vessel would require slipping for examination of the underwater parts and fastenings, renewal of deck and hull planking at the aft part, repairs or part renewal of main frames, refitting of

stringers and knees, extensive caulking in hull planking and removal of side linings for internal examination of the for'ard hull structure. He estimated that this work would cost over £7000, excluding any repairs found necessary below the waterline and in the for'ard part of the vessel.

He advised against lifting the *Swan* out of the water in its current condition for transport by road or onboard a ship without first slipping the vessel for examination and repairs. He concluded: "It appears to us therefore that restoration of the vessel to her original condition as a sailing fishing drifter may be considered an uneconomical proposition."

Having studied the report the members of the Shetland Maritime Trust decided that restoration of the *Swan* was unfeasible and turned their attention to another boat the *Research* which they later donated to the Scottish Maritime Fisheries Museum at Anstruther.

The *Swan* changed hands again and was brought to Hartlepool, where her new owner carried out alterations prior to taking the vessel to Spain. The first attempt ended in failure when the *Swan* suffered engine trouble and had to return to Hartlepool. On the second attempt the *Swan* ran into bad weather and had to be escorted into Whitby by the local lifeboat. She was then taken back to Hartlepool, the plan to take her to Spain being abandoned.

That might have been the end of the *Swan* had she not been spotted by John Fisher, a businessman from York who describes her condition at that time:

I first met Swan on an overcast August day in 1974. She was lying forgotten and derelict in Alexandra Dock, Grimsby. All of the decks, deck beams, beam shelves and rotten frame heads had been removed. Fifty per cent of the deck had been replaced but not caulked; and large areas of the hull planking were missing. Even in this sorry state Swan had a presence and spirit which I believe never left her.

I found the owner Arthur Odell in September 1974 and paid £400, which was to include fittings and timber to complete the hull rebuild. Unfortunately the old Caterpillar engine was frost damaged beyond repair and I therefore had to tow Swan to York where I lived at that time. As I stood on the deck, steering the boat by a jury rigged tiller, my abiding thought was "What have I done?"

There followed four years of non-stop hard work during which time I received a great deal of help, particularly from Stuart Hebden of the Hebden boatyard and from a local engineer, Philip Crowder. The old Cat 2U engine was

replaced with a rebuilt engine from the American army. The wheelhouse and much of the deck equipment came from Grimsby vessels. Much of the interior work was done by Hebden whilst Crowder Engineering were involved in all of the mechanical works.

In July 1978 Swan was slipped at the New Holland shipyard and a full survey carried out by Brodrick Wright and Strong of Grimsby (this survey is now held by the Swan Trust). The survey promised her to be in good condition and once again fit for sea. During the following four years I lived onboard and cruised extensively, going as far as Den Helder in northern Holland, and was a frequent visitor to many East Coast ports. Swan was a good passage maker and would always average nine knots.

In 1982 I took Swan to Hartlepool where I sold her to a local man, Tom Southern. The sale price was £22,000 which is an indication of her condition. It was therefore surprising that he abandoned her.

She was laid up in Hartlepool for several years, steadily deteriorating and sinking two or three times due to neglect. In 1989 she attracted the attention of local businessman Keith Parkes, who realised that the old boat, lying submerged with only her masts showing, was "a classic ship" and bought her without even knowing her condition.

She was lifted out of the dock by crane and only then did her true state become apparent. She had been lying in a tidal dock and her port side had been damaged through contact with the quay. She was filled with 15 tonnes of concrete ballast, all of which had to be removed, revealing several broken timbers.

On 9th March, 1990, when *The Shetland Times* carried a report on the rediscovery of the *Swan*, Mr Parkes had been working on her for four months. He had given her a lot of new planking and her old engine had been removed, being replaced with a 240 hp Cummins diesel engine. Mr Parkes told *The*

Shetland Times that he intended to bring the *Swan* back to Lerwick later that year.

The job took far longer than Mr Parkes had expected and the planned visit to Lerwick did not materialise. In its issue of 28th September, 1990, *The Shetland Times* reported that Mr Parkes had decided to offer the *Swan* for sale — "very reluctantly because of business commitments." He made it clear that he would seek no more than the cost of purchasing her and the work he had carried out.

The Shetland Times responded with enthusiasm, describing this as "a rare opportunity to retrieve part of Shetland's maritime heritage."

This suggestion was taken up the following week by Thomas Moncrieff of Lerwick, a retired technical and navigation teacher and keen yachtsman, as well as an enthusiast for all things pertaining to Shetland's maritime heritage. In a letter to the editor of the newspaper he wrote:

A Mr Keith Parkes has had the Swan lifted from the bottom of Hartlepool dock, restored so as to be fit to go to sea again. Most ports all round Britain now have maritime museums of some kind, but Shetland is lagging sadly behind ... a few weeks ago the Shetland Amenity Trust was willing to spend nearly £10,000 to break up the controversial remains of the Jessie Sinclair in Effirth. If three and a half times that sum cannot be obtained to buy the Swan, then Shetlanders need not talk of boat museums and maritime heritages. There will never be another Swan.

His reference to the *Jessie Sinclair* recalled another failure on the part of Shetlanders to preserve a historic fishing vessel. She was one of the MFV class, built for Admirality duties during World War Two with fishing in mind when peace should be restored. On one memorable trip to the fishing grounds at East Anglia she won the Prunier Trophy for the top catch of the season — the only Shetland boat ever to win this coveted prize.

THE SWAN TRUST

Mr Moncrieff's letter caused a great deal of interest in Lerwick, the town where the *Swan* had been built, and in Whalsay, the island where she had spent most of her working life. The man who brought all the interested parties together was James Moncrieff — Tom Moncrieff's son — who was then the chief executive

of Shetland Salmon Farmers' Association. He called those interested to a meeting in the association's offices at 80 Commercial Street, Lerwick. It was attended by eleven people who formed themselves into the Swan Steering Group with James Moncrieff as chairman/treasurer and Vaila Wishart as secretary. The others were Ileen Anderson, William Anderson, Alistair Hamilton, Dave Hammond, Tom Moncrieff, John Ratter, Allan Wishart, Brian Wishart and Robert Wishart.

It was pointed out that the *Swan* had been on the market for two months at an asking price of £36,000 exclusive of VAT and it was agreed that unless some action was taken fairly quickly the opportunity to purchase the *Swan* might be lost.

The general feeling was that if it proved feasible to bring the *Swan* back to Shetland she should become a "working boat" and not simply a static "museum piece."

It was agreed that a professional survey should be undertaken on the hull to determine the feasibility of the project and that a letter should be sent to Shetland Amenity Trust requesting initial funding of a maximum of £750, to cover the cost of this initial survey. If the application was successful Mr Laurence Irvine of the engineering firm Malakoff & Moore should be asked to undertake the survey on behalf of the group.

Mr Parkes wrote to Mr Moncrieff detailing the investment he had made in the vessel. He pointed out that the main reason for offering her for sale was that he and his wife intended to move north to Caithness to buy a hotel there.

Mr Parkes enclosed a report by Captain J. L. Elliott, a marine surveyor, who confirmed that he had inspected the vessel weekly while repairs to the hull of the vessel were being carried out and had witnessed the purchase of 688 feet of inch thick planking and the replacement of some 650 feet of planking around her hull. He added the comment: "Mr Parkes is to be congratulated on the work he has done to date."

The steering group appointed Laurence Irvine to carry out an inspection on their behalf. This was done at Jackson Dock, Hartlepool on 27th and 28th November, 1990. Mr Irvine was then retired after a long career as a carpenter, then surveyor, with Malakoff and Moore Ltd and had an unrivalled knowledge of wooden fishing vessels.

In his report Mr Irvine described how the vessel had been given a new wheelhouse "typical of wheelhouses seen in Danish fishing vessels", with sides

48. Under sail again at last. The *Swan* returns to Whalsay, 1996. Photo: Dennis Geldard.

49. A wave from Tom Moncrieff as he savours the fruits of all his hard work for the Swan Trust. Tom was instrumental in the formation of the trust and the restoration project. Photo: Dennis Geldard.

THE SWAN

50. A herring smack under sail once more in Shetland waters. The *Swan* has become a familiar sight locally and is in constant demand in the summer months both around the isles and for trips further afield such as Norway and St. Kilda. Photo: Dennis Geldard.

51. Swan Trust chairman from its inception until 1998 Jimmy Moncrieff (left), his father Tom who sparked off the whole *Swan* project (at the wheel) and Andrew Tait preparing to gybe off Sumburgh. Photo: Allister Rendall.

52. *Swan* berthed at Ham, Foula, during her first cruise around Shetland, July 1997. Photo: Allister Rendall.

53. The old and the new. Nothing can give a better idea of the "progress" of the Shetland fleet through the century. The *Swan* berthed alongside the mighty *Altaire* at Collafirth, Northmavine, 1997. Photo: Tommy Watt.

54. The *Swan* off Watsness, the west side of Shetland, in 1998. Photo: Captain Malcolm Brannan.

55. Fine sailing – the massive mainsail provides the drive. Photo: Peter Robertson.

56. The first full-time skipper of the restored *Swan*, Andrew Halcrow, points out some landmarks for his young crew, 1998. Photo: David Marwick.

57. Crew member Sandra Hollingdale introduces trainees to some chartwork. Photo: Dougie Grant.

58. A trip around the spectacular seabird cliffs of Noss is a treat for these young sailors from Sound Youth Club, 1998. Photo: Dougie Grant.

59. There is always plenty of heaving and pulling on ropes onboard a sailing boat, especially a restored smack-rigged herring boat. Photo: Dougie Grant.

60. The infamous Brent Spar was a bit of a tourist attraction when the *Swan* visited Norway in 1998. Photo: Tommy Watt.

61. Norwegian holiday, 1998. Although not the best for hard-case sailors the Norwegian fjords have their own attractions on lazy summer days. Photo: Tommy Watt.

62. The *Swan* under full sail – jib, staysail, mainsail and "dandy". Photo: Kieran Murray.

63. Sound Youth Club members enjoy a night cruise (from left): Anne Wisdom, Janie Yeaman, Cheryl Stewart and Megan Gair. Photo: Andrew Halcrow

64. Returning from Baltasound with the Shell Sailing Club, June, 1999. Photo: Andrew Halcrow

65. Cunningsburgh School pupils soaking up the sun on a trip around Bressay. Photo: Andrew Halcrow

66. A comfortable berth! Photo: Andrew Halcrow

67. Landfall at the north end of St Kilda, 1st June, 1999. Photo: Andrew Halcrow

68. The "St. Kilda crew" pose for a photograph off Skye. Photo: Andrew Halcrow

69. Repairs on the way to St. Malo, Richard Stafford cuts out a temporary glass for the binnacle. Photo: Andrew Halcrow

70. Zander Simpson takes a sun sight. Photo: Andrew Halcrow

71. Ingrid Eunson and Richard Stafford stand a night watch. Photo: Andrew Halcrow

72. Alongside at St. Malo Fiddlers' Bid entertain the crew and onlookers. Photo: Andrew Halcrow

73. Visitors from a local yacht club in St. Malo pay a visit to the *Swan*. Photo: Andrew Halcrow

74. The *Swan* leaving St. Malo. Photo: Andrew Halcrow

75. The Tall Ships Race starts. Photo: Andrew Halcrow

76. At sea – members of Fiddlers' Bid in the galley. Photo by Andrew Halcrow

77. The *Swan* underway in the Irish Sea. Photo: Andrew Halcrow

78. Harpist Catriona McKay accompanies Don Farquhar (on the spoons and wheel!). Photo: Andrew Halcrow

79. The *Swan* heading for the start of the Lerwick to Aalborg leg of the Tall Ships Race, 12th August, 1999. Photo: Keiran Murray

of hardwood tongue and groove boards and roof of laid planks sheathed with glass fibre. The existing wheelhouse was being extended aft by 7 feet 6 inches.

He noted some decay and softening of the timber along the base of the existing wheelhouse and added "otherwise the structure generally appears satisfactory with the windows intact and undamaged."

The original timber heads had been cut off level with the tops of the deck beams and a wide margin plank formed the perimeter of the deck. The decks were of softwood or red pine laid and caulked — not original. Several places were noted as having been renewed "recently".

Apart from the repair planks all being butted on one beam, the decks "generally appeared to be in reasonable condition." However "numerous isolated soft spots were found by prodding." His examination of the deck beams showed minor areas of decay in some of the original ones.

Turning to the interior of the vessel Mr Irvine found decayed upper sections in some of the frames. In the bilges the frames showed "slight superficial wasting where the timber has been more or less permanently wetted during the life of the vessel. Otherwise the frames were hard and as far as could be ascertained in sound condition."

He concluded:

The examination of the structure has not brought to light any significant defects on the original parts which were seen, apart from the fore deck beams. The decay found does not appear to have reached the stage where all the forward beams would have to be discarded in a restoration project. The deck planking (not original) would require 100 per cent renewal but with care and coating it could serve for some time yet.

The underwater hull, which was not seen, was reported by Capt. Elliott to have been refastened and a total of 688 linear feet of strakes renewed.

On this basis I assume the underwater parts to be in satisfactory condition and I can see no reason providing the vessel is regularly maintained why Swan should not serve as an example of Shetland's fishing and boatbuilding heritage for many years to come.

On the basis of this information it was agreed to form the Swan Trust. At a meeting of the Swan Steering Group on Thursday 13th December, 1990, it was agreed to write a letter to Mr Parkes confirming the group's interest in the *Swan* and its intention to proceed with the purchase of the vessel. Robert Wishart

agreed to produce a brochure with photographs of the *Swan* for fund raising purposes. It was anticipated that the work of restoration would cost a total of £150,000, being divided into three stages — purchase and travel, initial restoration and restoration to her original rig as a sailing smack. It was assumed that each phase would cost around £50,000.

The final meeting of the Swan Steering Group was held on Thursday, 15th December, 1990, followed later that evening by the inaugural meeting of The Swan Trust.

The membership of this body included several of those who had been involved in the project ever since discussion began on the possibility of bringing the vessel back to Shetland. They included Robert Wishart, Vaila Wishart and Thomas Moncrieff, all of Lerwick, William Simpson and William Anderson of Whalsay and Brian Wishart of Sandwick.

In addition it was agreed that the trust deed should specify representation by several groups, including community councils and boating clubs in Lerwick and Whalsay. Allister Rendall was nominated by Lerwick Community Council while John L. Simpson represented Whalsay and Skerries Community Council. Alistair Hamilton represented Shetland Amenity Trust, Thomas Watt, Shetland Museum and Lindsay Aitken was nominated by Lerwick Harbour Trust.

Purse seiner skipper David Hutchison was nominated by Shetland Fishermen's Association; ex-skipper James H. Henry represented Shetland Fish Processors Association; and John Ratter was appointed to represent Shetland Salmon Farmers Association. James Moncrieff, originally nominated by Shetland Salmon Farmers Association, became a trustee in his own right following his resignation from the post of chief executive of the association.

Under the trust deed the trustees were given wide ranging power to appeal for and receive donations, grants and legacies and to raise money "by any other means which they may consider appropriate for the purpose of accumulating and maintaining funds for the purpose of the trust."

They were authorised to open bank accounts which should be operated upon by such persons and in such manner as the trustees might from time to time direct.

The objects and purposes for which the trust was established were described as "generally of an educational nature", the aim being to:

... acquire the hull of the Shetland Fifie Swan (LK 243): to restore and re-rig the Swan as a working and sea-going sail-fishing boat using as far as reasonably

practicable similar materials and techniques to those used in her original construction and design; to ensure the continued maintenance of the Swan in her restored and re-rigged condition as aforesaid so that she may be preserved and open to the public as part of Shetland's maritime and fishing heritage for future generations; and to encourage and facilitate interested parties and particularly young people to sail on the Swan, thereby teaching and keeping alive the techniques of sailing and working a traditional sail Fifie such as the Swan.

Negotiations to purchase the vessel were concluded fairly quickly. She cost £41,525 including VAT — a considerable sum for a vessel of her age, but then she was no ordinary vessel. All those involved in the project realised that this was the last chance to acquire a Shetland-built sail fishing vessel.

Funding for the first part of the project, the purchase and retrieval of the *Swan* was soon forthcoming as Shetland Amenity Trust gave a grant of £30,000 and other funds came from Lerwick Community Council (£7000), Shetland Salmon Farmers Association and Shetland Fish Processors Association (£5000 each), Shetland Fishermens' Association (£4000) and Lerwick Harbour Trust (£3000).

From the membership of the trust a "Swan Return Group" was formed to make arrangements for bringing the vessel back to Lerwick. This involved a great deal of careful planning, since the vessel had to be seaworthy and had to satisfy the requirements of the insurers through Captain Elliott before she was allowed to undertake the long journey north.

A lengthy list of equipment was prepared by Dennis Geldard — pumps, flares, anchor chains, liferafts as well as the more mundane pots and pans and a kettle for the galley. Some of the more expensive items were loaned by people or firms interested in the project. Two liferafts came from B.P., an emergency radio beacon was loaned by Dave Hammond and Tom Moncrieff agreed to lend a staysail which had been used originally on another sail boat *Gracie Brown* and also on the ex-fishing boat *Loki* from 1951 to 1971.

This equipment was taken south free of charge on Hay & Company's vessel *Shetland Trader* — an appropriate gesture from the firm which had built the *Swan* and which is still thriving as a subsidiary of John Fleming & Company of Aberdeen.

The delivery crew left Lerwick on Friday 5th April, 1991. They were fortunate in having as skipper William Simpson who, many years before, had

served on the *Swan* and whose family had owned her for most of her career as a fishing vessel. Engineer was George Sinclair of Trondra, loaned to the trust by his employers Malakoff & Moore because of his skill with engines. His knowledge was to be tested to the full on the way north. The other two crew were Tom Moncrieff and Dennis Geldard.

On arrival at Hartlepool they began to prepare the *Swan* for her journey. The electronic equipment which they installed included a radar, an echosounder and a Decca navigational system — instruments which hadn't even been dreamed of when the *Swan* was fishing, and which demonstrated the enormouos technological advances of the 1950s.

The hull was entirely empty and the first task undertaken by her crew was to lay temporary flooring to accommodate 17.5 tonnes of ballast considered necessary for stability. There was no sleeping accommodation, so a temporary bunk, big enough for three, was constructed for'ard of the hold. George Sinclair tested the engine for several hours and pronounced it fit for the journey.

After four days work the *Swan* was ready for her journey. The surveyor was satisfied that she was fit for the trip and Tom Moncrieff had completed the delicate job of swinging and adjusting the compass.

With full fuel tanks the *Swan* left Hartlepool at 12.15pm on Wednesday, 10th April and by 12.30 she was clear of the outer buoy. By 8.00pm she had the Longstone Light abeam. The wind was fresh and from the south-west, making ideal conditions for the journey. The *Swan* was retracing a journey which she had last made more than 80 years ago before.

During the night a fault developed in the electrical system as the generator "boiled" the acid out of the batteries. The choking gas could have asphyxiated the off duty men if George Sinclair had not gone into the engine-room and pulled off wires but this meant that they had no lights and put the radar was out of action.

Next morning at 10.00am they berthed in the fish dock at Aberdeen. The rest of the day was spent in overhauling some gear while an electrician carried out repairs. Two more crew members, Robert Wishart and Allister Rendall, joined the *Swan* at Aberdeen. They left there at 10.25 on Friday morning, heading into a fresh northerly wind. Next morning at 2.30am they had their first connection with home when they met the Aberdeen - Lerwick ferry heading south three miles to port. At 9.50am the *Swan* was abreast of Sumburgh Head and from there the engine was put on full speed for Lerwick.

Enthusiasts gathered at the Knab, a headland at the entrance to Lerwick harbour, and video cameras whirred as she came in past the Bressay lighthouse. The film company Cinecosse, who had filmed her departure from Hartlepool, captured the memorable last stage of the journey for their video which was then used to promote the *Swan's* restoration. At 12.30 the *Swan* tied up in the small boat harbour, after an absence from Lerwick of more than 30 years.

The *Swan* was the centre of attention during that summer's seafood festival in August. She was berthed at Albert Dock where visitors were invited to examine the vessel and to see an exibition of photos covering her unusual history. They also contributed over £200 to the restoration fund.

THE RESTORATION

The inaugural meeting of the *Swan* Restoration Committee was held on Tuesday 27th August, 1991. By this time the *Swan* had been hauled up on the Malakoff slipway, where a closer examination revealed that the repairs required would cost far more than expected.

It was clear that she would need a new keel and new bilge stringers, while all her decks would have to be renewed. It was also noted that much of the planking carried out at Hartlepool had been done without sufficient care being taken over the positioning of the butts and it was clear that much of this would have to be re-done.

The new estimate for the cost of the repairs under Phase Two was now put at £100,000 - double the original estimate. Fortunately the charitable trust of Shetland Islands Council responded with the offer of £75,000 to cover half the expected cost of restoration.

Previous surveys had shown that the *Swan* needed a new keel and keel shoe and that parts of the bilge stringers would have to be replaced, as well as some of the strakes. It was agreed that the high rail and iron stanchions added at Grimsby would have to be replaced by the low gunwale characteristic of herring boats in the early part of this century. This would involve the replacement of all 88 timber heads. Finally the vessel would require a new deck after 12 deck beams and 16 part beams had been replaced.

The Swan Trust approached several boatyards for estimates to carry out the repairs; but in the end it was decided to award the contract to the local shipyard Malakoff & Moore. The yard would carry out the repairs to the hull at Lerwick

and overhaul the engine at its Scalloway premises. The remainder of the restoration work, which included the renewal of the timber heads and re-decking, would be carried out under a working relationship with the trust.

It was agreed that the *Swan* would employ Wilfred Bruce and Gordon Smith to work directly on their behalf with Billy Smith of the Malakoff as foreman. It was also agreed that Tom Moncrieff should oversee the entire project on behalf of the trust.

The *Swan* was lifted ashore at Lerwick's Morrison Dock by a crane hired from Shetland Line, capable of lifting up to 35 tonnes. By this time the *Swan's* wheelhouse, engine and ballast had been removed to reduce the weight to be lifted.

It was when the trust's employees were cutting the tops off the frames, to get to sound wood prior to replacement, that the true state of the framing became apparent. While an earlier survey had indicated that the frames were in good condition it now became clear that only three of the 44 frames were partly sound. The carpenters drew the trust's attention to this disastrous state of affairs, the news being met at first with disbelief.

Since it was assumed that the planking was generally sound, it was agreed that the frames should be removed one by one and replacements cut to fit the shape of the hull at each point from templates.

The carpenters had an advantage over their predecessors who had built the *Swan* in the form of an electric chainsaw to cut out the parts of the frames. The parts of the original frames which were sound were incorporated into the new structure.

This job had not gone very far when there was some more bad news to report to the trust, when it was discovered that the inner side of the planking was generally in a poor condition and that the entire planking would have to be replaced.

Those who have studied the problem in detail have suggested that one explanation for the unforeseen poor state of the hull was the way that the nails had become oxidised, allowing water to penetrate into the planking, resulting in rot between the planking and the frames. It is suspected that the long period when the *Swan* lay sunk in brackish water, laced with chemicals, had a serious effect on the nails.

There was also concern over the new planks which had been added at Hartlepool. Many of them were short and the butts of adjacent lengths were

often too close together to meet the specifications laid down as requirements for a vessel used in sail training. It had also been the trust's intention that the boat would be used to instruct a new generation of Shetlanders — and visitors — in handling a large vessel under sail. The decision to replank the vessel ensured that there would be no problems when the time came to apply for certification.

Rebuilding of the *Swan* went ahead throughout the summer of 1992 and in November, after the carpenters had been working under atrocious conditions for some weeks, it was decided to erect scaffolding so that the vessel could be enclosed to allow work to continue throughout the winter.

What had started off as extensive repairs had become a rebuilding to the highest possible standards for sail training. Apart from the small pieces of the frames one of the few parts of the original vessel to be retained is the keelson, or hog as it is known in Shetland. It is of pitch pine and it is as good as new.

She was given a new fore stem and a new apron was fitted, this painstaking job being entrusted to carpenter Raymond Sinclair. The old apron bore evidence of her collision with the cliffs of Bressay. Having been split, it had been cross-bolted while repairs were being carried out by a previous generation of Malakoff workmen. She was also given a new after stem and an apron down as far as the deadwood — another part of the original vessel to be retained.

The main keel was shaped from two oak trees, the parts scarfed together. The frames are also of oak — 4.5 inches thick and 9 inches deep at spacings of 13 inches between the frames. Bilge and deck stringers, which run the entire length of the vessel to consolidate the frames, are heavy lengths of oak 3 inches thick. The two welts or rubbing strakes are also of oak, 3 inches thick and 4 inches wide.

The deck beams are of larch, 7 inches deep and 6 inches wide, attached to alternate frames. Hatch coamings are also of larch, 4 inches thick and averaging 15 inches deep. The decks are made of iroko, the planks 2.5 inches thick and 5 inches wide. They are caulked with cotton thread and the spaces sealed with Sikaflex.

Care was taken to re-create the special arrangement for lowering the mainmast. This stands in a massive tabernacle, pivoted 4.5 feet above deck level. When standing upright it is fastened securely at the foot of the tabernacle. To allow the mast to be lowered a 12 foot long trunk or slot in the deck allows the heel of the mast to swing forward until it protrudes above the deck.

From the start of the restoration the trust intended that the *Swan* would be a

working vessel; and it was therefore necesssary to meet the stringent standards required by the Department for Transport for small commercial sailing vessels. Four steel bulkheads have been fitted, complete with independent pumping systems. One of the bulkheads is near the bow to help reduce the effect of a collision.

As reconstruction went ahead it became clear that the original funding was inadequate. While the charitable trust of the Shetland Islands Council agreed eventually to provide a further grant of £125,000, there was inevitably a great deal of reluctance on the part of some councillors to do so and they asked some very searching questions as to why the original estimates had been so far out, why the work was taking so long and whether the trust had considered how the vessel would be funded when the work was complete.

This was an anxious time for the trustees. The publication in *The Shetland Times* of a cartoon by "Smirk", showing a swan on its crutches, helped to lift the gloom at a critical time. Eventually the grant was cleared by the full council, allowing the second stage of the *Swan's* restoration to be completed.

The restoration committee received valuable help from several people including John Ratter, and Jack Duncan of the staff of Malakoff & Moore supervised the planking. Sadly the trust was denied the experience of Laurence Irvine, who died in September 1992 — a great loss to the trust, which had relied so much on his expertise.

A bonus for Malakoff & Moore was that the work provided valuable experience for several of its staff who seldom nowadays will have the opportunity to assist in the building of a wooden vessel. Raymond Sinclair, Sidney Sinclair, Jim Tait, Jackie Priest, Tommy Duncan and Charles Peterson all had an opportuntity to demonstrate their skill, while apprentices such as Angus McNeil from Yell, Douglas Moore from Lerwick and John Martin Tulloch of Sandwick have gained much experience, which ensures that the skill of building and repairing large wooden vessels will continue into the next generation of craftsmen. In addition to this several of the Malakoff yard's skilled welders did excellent work, often working under difficult conditions.

After many months of hard labour came the day when the *Swan* was ready to take to water. A large crowd gathered at Morrison Dock on Saturday,16th April, 1994 to watch as the vessel was swung from the quayside by a heavy lift crane and lowered into the harbour. Mr Ian Munro, superintendent of the Fishermen's Mission at Lerwick, gave a service of rededication.

The *Swan* made a splendid sight with her distinctive colour scheme — bright red bottom and dark green topsides, separated by a white waterline which widened to a broad "shark's mouth" at the stem. Her low rails are painted blue and white and the decks are coated with a mixture of boiled oil and real turpentine, which both protects and enhances the natural appearance of the wood.

The *Swan* was taken to the small boat harbour at Lerwick where during the summer of 1994, she attracted a lot of attention from local people and visitors. Her small size contrasted markedly with the huge bulk of modern herring vessels — large purse seiners and pelagic trawlers with greatly increased catching power — as they lay at Victoria Pier or the outer arm of the breakwater, waiting their turn to go out to the klondykers to discharge their catches. Today these are measured in tonnes — not crans.

Another step in the *Swan's* restoration came in September 1994 when the Cummins diesel engine which had been installed at Hartlepool was replaced on a new steel engine bed, after being thoroughly overhauled by Malakoff & Moore. The firm also made a new propeller shaft and stern tube which had been installed before the engine was ready. The engine controls are housed on deck near the steering wheel near the stern.

The final stage in the *Swan's* restoration was to furnish her with masts and sails. The trustees were fortunate in that a scale model of the vessel in her original sailing rig is on show at Liverpool Maritime Museum. Mr Adrian Osler, author of the highly respected book *The Shetland Boat* and a strong supporter of the *Swan* project, obtained access to this model. He then produced photgraphs and drawings to indicate the position of the original deck fittings.

Based on this information and from knowledge gathered from the former sailing drifter *Gracie Brown*, Thomas Moncrieff produced a sail plan and worked out the dimensions of the spars, to restore the vessel to her original state as a smack with mainsail, mizzen, foresail and jib.

Masts and spars were made to the trust's specifications by T. N. Nielson of Gloucester from Oregon pine, after selecting an 80 year old tree growing in the Forest of Dean. Lerwick Community Council gave a donation of £8500 towards the cost of masts and spars and Hay & Company, the builders of the *Swan* back in 1900, agreed to transport the masts and spars free of charge in their cargo vessel *Shetland Trader*, provided that the supplier could deliver them to Berwick where the ship was due to call.

An essential part of the traditional herring smack was the steam capstan, used for hauling the bush rope to which the driftnets were attached, for raising the mast and for discharging the catch. The *Swan's* capstan had disappeared many years before her restoration but it came to the attention of the trust that the capstan of another of the *Swan*'s contemporaries, the *Laurel,* which had been broken up at Cunningsburgh in the 1950s, was lying above the beach there. Mrs Joann Halcrow, whose husband had bought the *Laurel* for breaking up, agreed to give it to the trust.

Another Cunningsburgh man Pete Adamson (later to lose his life while on a trip to Albania on behalf of Shetland Aid Trust) collected it from the beach free of charge.

The capstan was taken to the workshop of HNP Engineers in Lerwick to be converted to hydraulic power before being installed in the *Swan*, aft on the starboard side, the position favoured by the crews of the Shetland sail boats.

The engineers found that all of the essential parts were intact but the whelps, the alternating blocks of wood and iron attached to the central stalk, were missing. It wasn't difficult to make replicas of the wooden whelps but the loss of the iron whelps was a more serious problem.

It was then discovered that the original iron whelps from the *Swan's* capstan had been donated to the Scottish Fisheries Museum at Anstruther. The trustees of the museum agreed to lend one of the whelps to the Swan Trust so that castings could be made.

The *Swan's* first journey under sail after her restoration was to Whalsay in the first week of June 1996. There she was reunited with several men who had served on her as crewmembers when she was a motor fishing vessel. She then returned to Lerwick before setting out on a journey to Scalloway by the northerly route, arriving there on Saturday 15th June. After having her bottom repainted on the slipway there she headed west to Walls to accompany five yachts in the round Foula race on Saturday 22nd July. She spent the night at the Walls pier when a fresh crew joined her to take her back to Lerwick via Sumburgh Head.

Among the visitors to see her after she returned to the small boat harbour was John Botting, who owned her while she was at Grimsby and had maintained a close interest in the restoration after seeing, in a copy of *Fishing News*, a photograph of the *Swan* being lifted back into the sea.

During most of 1996 work had been progressing on the internal furnishings

of the vessel, designed to equip her for her new role as a sailing training vessel. A companionway from the after hatch leads down to the cabin which, due to the engine room, is slightly smaller than the original. It has seven bunks and lockers for personal belongings.

A doorway through a watertight steel bulkhead leads from the cabin into the engine room, with its Cummins engine, a 10KW generator and battery charger. It also houses two fuel tanks and the hydraulic system for the capstan. There is also a small toilet. The for'ard bulkhead of the engine room is another steel structure. On the other side of the bulkhead, with access from another hatch on deck, the space formerly occupied by the hold or fish room is the main accommodation, with eight bunks, small microwave cookers, for quick snacks during a voyage, and two small toilets. Here too are a VHF radio and a satellite navigation system. Much of the space here is occupied by a large table for chart work and for meals.

Farther for'ard, through another watertight steel bulkhead, is the main galley with two fresh water tanks, another two toilets, a washbasin and a shower. Here too is the workshop, gear store and chain locker. Near the stem is a steel collision bulkhead.

The *Swan* has several other pieces of equipment unheard of when she first sailed to the fishing grounds. Tucked out of sight under the hatch is a console with a Koden radar, a Furuno echosounder and VHF radio set. Near the stern are the engine controls and the steering wheel with chains leading to the rudder.

The complicated electrical system was installed by John Arthur Shearer of Clate, Whalsay, a crewman on a modern fishing boat.

SWAN IN SERVICE

A full time skipper was appointed for the *Swan* in December 1997. From a large number of applicants Andrew Halcrow from Hamnavoe, Burra was considered most suitable for the job.

None of the other applicants could equal his record of seamanship under sail. With his brother Terry he had built a 32 foot long steel-hulled yacht named *Elsi Arrub*, designed for her sea-keeping qualities rather than speed. They left Hamnavoe on the 16th June, 1988, and in the next five years they circumnavigated the world, with lengthy stops in St. Martin, Auckland, Melbourne and Sri Lanka.

When skipper Halcrow took command of the *Swan* she had a certificate

valid for a year which enabled her to carry passengers. A more extensive survey was carried out which resulted in the granting of a "small commercial vessel" certificate valid for five years.

The *Swan* had a busy season in 1998. She carried nearly one thousand trainees and sailed more than three thousand nautical miles. In addition she had more than a thousand visitors on her open days. The season started on the 12th April with a short trip south of Lerwick, followed by a journey "round the heads" to Scalloway, where she was taken up on the slipway to be repainted. Thereafter she was kept busy taking parties of schoolchildren, youth clubs, senior citizens, etc. on what for many of them was the journey of a lifetime, recalling the great days of Shetland's sail fishermen. While most of these trips started and ended in Lerwick the *Swan* also spent several days in the Brae area, sailing with parties of schoolchildren in the St. Magnus Bay area.

Early in June, with a party of schoolchildren, she set off on a week's visit to Norway, calling at several ports in the Bergen area. After leaving Leirvik bound for Lerwick she encountered severe weather. All went well until the vessel was a few hours away from Lerwick when the steering chain on the port side snapped.

Skipper Halcrow decided to contact Shetland Coastguard who in turn requested that Lerwick lifeboat should go to the *Swan's* assistance. The lifeboat reached the *Swan* about 4 am and four and a half hours later she was back in her berth in the small boat harbour, none the worse for her ordeal. The faulty chain was replaced by a stronger one.

After more short trips from Lerwick the *Swan* sailed again for Norway on the 23rd June reaching Måløy two days later. There was another trip to Norway

in July with a party of pupils from the Anderson High School. There is little doubt that the *Swan* is strengthening links between groups in Shetland and their counterparts in Norway.

A high point of the year was the trip to Arbroath for the town's Seafest in mid-August when 800 people visited the *Swan* during those two days.

She attracted just as much attention on her first long trip of 1999 when she set off for St. Kilda, dropping anchor in Village Bay on the last day in March. Her homeward journey was punctuated with stops at Leverburgh in Harris, Portree, Stronsay and Kirkwall.

The biggest event in her new career came in July 1999 when the *Swan* participated in the Cutty Sark Tall Ships race, in which Lerwick was a host port for the first time. With a crew of 15, consisting of young people and experienced sailors, the *Swan* left Lerwick on the 11th July, bound for St Malo in France, where the first leg of the race was started on the 24th of July. Among her crew were members of the local group Fiddlers Bid, who enlivened proceedings throughout the race.

Sailing conditions were reasonably good between St Malo and the Scilly Isles. Then as the *Swan* headed north the wind freshened from the north-east, impeding the vessel's progress considerably as she tacked up the Irish Sea. Around midnight on the 26th July the tack hook on the bowsprit straightened out with the force of the wind and all hands were called on deck to take down the sail.

Skipper Halcrow realised that the vessel would not reach Greenock within the time limit and decided to head for the Irish port of Crosshaven to take on more fuel. There was a bonus here for Fiddlers Bid when they were invited to play at a birthday party.

The *Swan* left Crosshaven at 4 am on the 28th July, heading for Greenock in light winds, having to motor most of the way. Light wind had affected all of the ships in the race and only half of the fleet was there when the *Swan* arrived.

The second leg from Greenock to Lerwick started of the 2nd August, with one additional crewmember - an Italian from the *Corsaro II*. Again head winds delayed the *Swan's* progress and she spent a night in Oban.

On the afternoon of the 4th August the *Swan* rendezvoused with the *Spirit of Scotland* at Sanna Bay, where the two crews enjoyed a barbecue. Then it was back to business as around 11 pm the crews hoisted sail and headed for the next stop at Stornoway, arriving there early on the 5th August.

There had been another delay when the *Swan's* propeller had picked up a length of discarded rope, which caused a reduction in speed. The problem was rectified when the mate, Terry Halcrow, assumed the role of diver and cleared the obstruction with a knife.

After a stay of only a few hours at Stornoway the *Swan* started the next leg of her journey to Baltasound, a planned stopover on the way to Lerwick. At Baltasound she joined a number of ships, their crews taking part in a programme of events and excursions organised by local people.

The *Swan* left Baltasound at 4 pm on Monday the 9th August, and arrived in Lerwick at 10.30 am. The next few days provided a week that Lerwick will never forget with more than 60 sailing ships in port, berthed all along the waterfront from the boating club to Shearer's Quay. Sailors from many different countries added their own distinctive atmosphere to Commercial Street and more than 5000 visitors from outwith Shetland joined local people in the lavish entertainment along the waterfront. The *Swan* was given a prominent berth at Victoria Pier in the company of some of the larger vessels in the race.

Then came the high point of the race as far as Shetland was concerned when the little *Swan* led the 'Parade of Sail' out past the Knab to the assembly point south of Bressay, for the start of the final leg of the race to Aalborg in Denmark.

This was a spectacle which those who witnessed will never forget, as the sea for several miles was covered with sailing ships, all waiting for the signal to proceed.

When the race was over the *Swan* headed back to Lerwick having completed another unusual chapter in her career.

Then it was back to the routine which she and her crew have established, instilling in a generation brought up in an age of diesel power a respect for the more arduous days of sail and providing the exhilaration felt when a vessel of this type is running before the wind.

When the plans to restore the *Swan* to her original condition were first announced there were sceptics who predicted that the exercise was futile and a waste of money. Events have proved them wrong. The *Swan* has attracted an enormous amount of interest everywhere she goes and sparked off a great deal of interest and pride in Shetland's rich nautical heritage in the long era of sail.

APPENDIX I

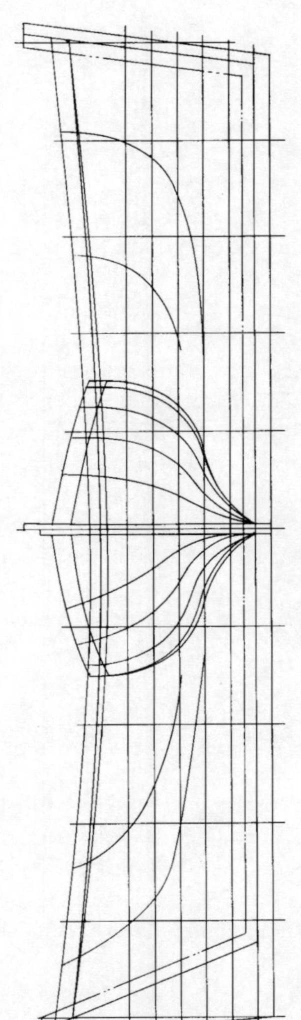

The *Swan*
Length overall: 67'10"
Length keel: 60'4"
Beam: 19'8"
Draft: 9'
Gross tonnage: 57

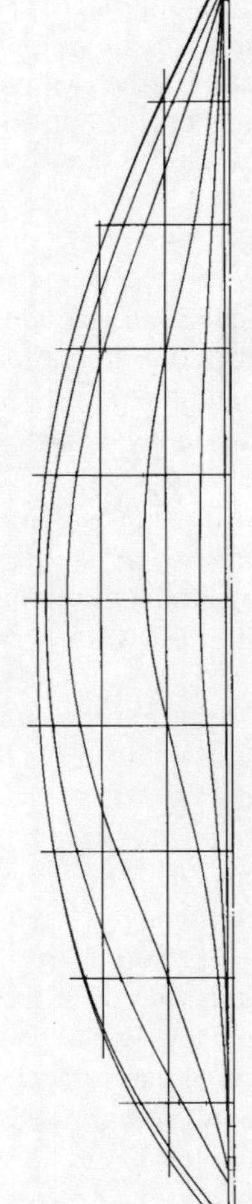

Lines drawing of the *Swan*. Courtesy of Angus Graham.

APPENDIX II

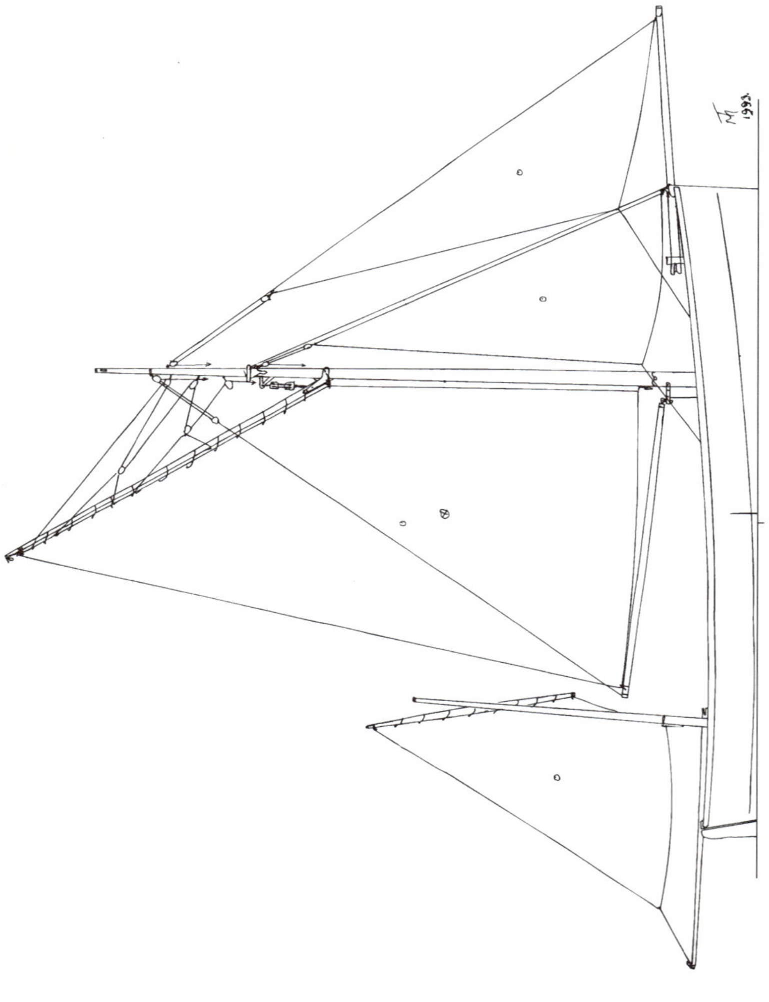

Sail plan. Courtesy of Tom Moncrieff.